はしがき

この「オートバイ＆電動車椅子のファイバースコープ バックモニター図形ラインで安全走行」では、従来の二輪車のようにハンドルに取り付けられたバックミラーの確認では見づらいこともあり、オートバイが走行中、左右・後方からの接触で事故に至るケースがあった。これを防止する知らせを予知することが困難であった。しかし、このバックモニター図形ラインでは、オートバイの運転者が容易に予知できることで接触事故を未然に防ぐことができるでしょう。

これからの高齢化社会における電動車椅子の需要は多くなり、歩道を走行する電動車椅子は、自転車や通行人との接触事故、また、ハンドルに取り付けられたバックミラーの確認では見づらいこともあり、左右・後方からの接触を予知することが不可能であり、特に高齢者による反射神経は、若年者と相違するから、これに対応するバックモニターで接触事故を未然に防ぐことができるでしょう。

Preface

This, ″Fiberscope of a motorcycle & electric wheelchair It runs safely with a back monitor figure line.″, then it was difficult to judge from confirmation of the rearview mirror installed in the handle like a conventional two-wheeled vehicle, and motorcycles were moving and contact from left and right and the back, and there was a case to an accident. It was difficult to foresee the news that this is prevented. But it's that a driver of a motorcycle can foresee easily by this back monitor figure line, and it would be possible to stop a fender-bender from happening.

Because the electric wheelchair by which an electric wheelchair in future aged society is in great demand, is and runs a sidewalk is difficult to judge from a fender-bender with a bicycle and a passerby and confirmation of the rearview mirror which was installed in the handle later, and it's impossible to foresee contact from left and right and the back, and a reflective nerve by a senior citizen differs with a youth in particular, it would be possible to stop a fender-bender from happening by the back monitor who answers to this.

目　次

第一章　二輪車、オートバイ、スクーター

１，ファイバースコープバックモニター図形ラインの解説---------------7

２、接触事故のパターン・追い越し・左折・右折・交差点(イラスト解説)-8

⑴　左側がらの接触　---8

⑵　右側がらの接触---11

第二章　電動車椅子

１，ファイバースコープバックモニター図形ライン解説----------------14

２，歩道の電柱と扉の間を通り抜け---------------------------------14

３，転倒・転落事故---15

４，歩道の障害物・電柱---16

５，横断歩道の転倒防止---17

英語解説

English description

Chapter 1 Two-wheeled Vehicle, Motorcycle, Scooter

1--18

Description of a fiberscope back monitor figure line

２---19

The pattern of a minor collision, passing, a crossing(illustration explanation)

(1)--19

Left-hand side contact

(2)--23

Right-hand side contact

Chapter 2 Electric Wheelchair

1, ---26

Description of a fiberscope back monitor figure line

2, ---26

It passes between the telegraph pole of a sidewalk, and a fence.

3, ---27

Accident of a fall and falling down

4, ---28

The obstacle and telegraph pole of a sidewal

5, ---29

It prevents breaking down from a sidewalk.

中国語解説

中文解说

第一章摩托车以及低座轻型摩托车

1, ---30

纤维镜背监视器图形线的解说

2、---31

接触事故的模式、超过、左转弯、右转弯、十字路口(插图解说)

⑴ --31

左侧的接触

⑵ --34

右侧的接触

第 2 章电动轮椅

１， --37

纤维镜背监视器图形线的解说

２， --37

在人行道的电线杆和屏之间穿过

３， --39

跌倒、坠落事故

４， --39

人行道的障碍物、电线杆

５， --40

人行横道的跌倒防止

ドイツ語解説

Deutscher Kommentar

Kapitel 1 zweirädriges Fahrzeug, Motorrad, Motorroller

１， --41

Der Kommentar des fiberscope kontrolliert zurück Zahl-Linie

２、 --42

第一章　二輪車、オートバイ、スクーター

１，ファイバースコープバックモニター図形ラインの解説

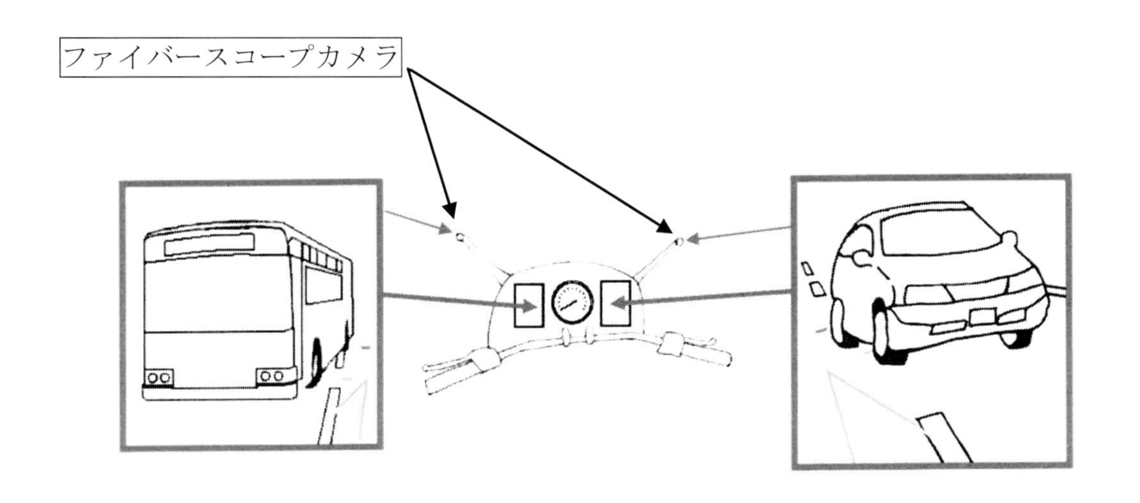

二輪車、オートバイ、スクーターのファイバースコープカメラはハンドルに取り付けられていた所に設置され、計器左右のバックモニターで見られる。

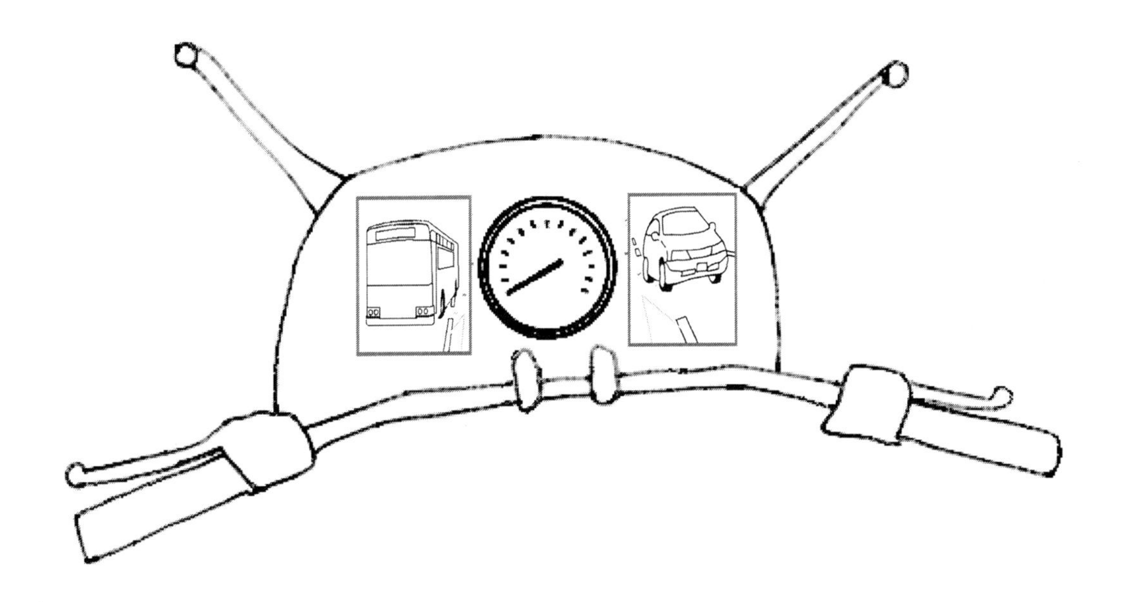

２、接触事故のパターン・追い越し・左折・右折・交差点（イラスト解説）

⑴　左側がらの接触

オートバイを追い越す車両が進路を変更することによる接触事故防止の確認

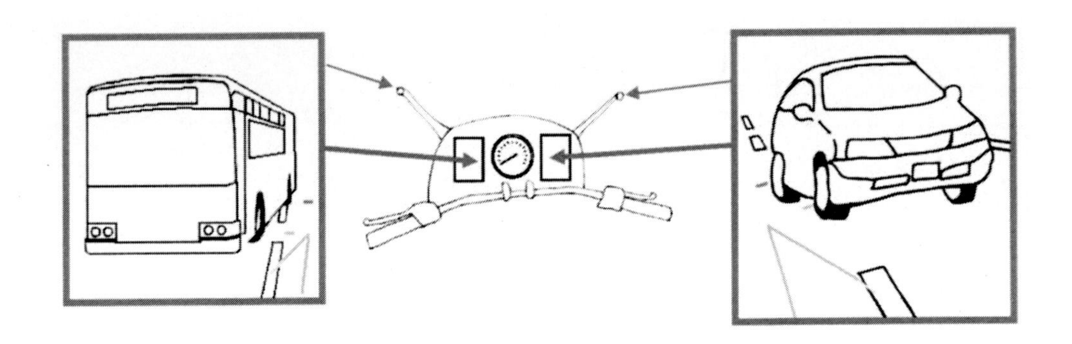

スピードメーターの左右にバックモニター、左後方を示す。

スピードメーターとタコメーターの左右にモニター、左後方を示す。

デジタルスピードメーターの左右にバックモニター、左後方を示す。

メーターパネル内の左右にバックモニター、左後方を示す。

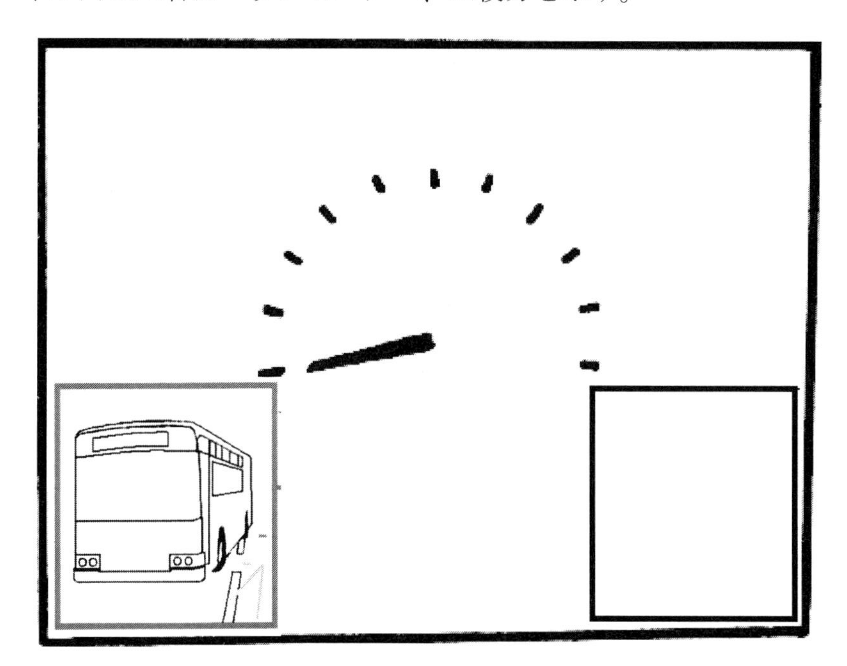

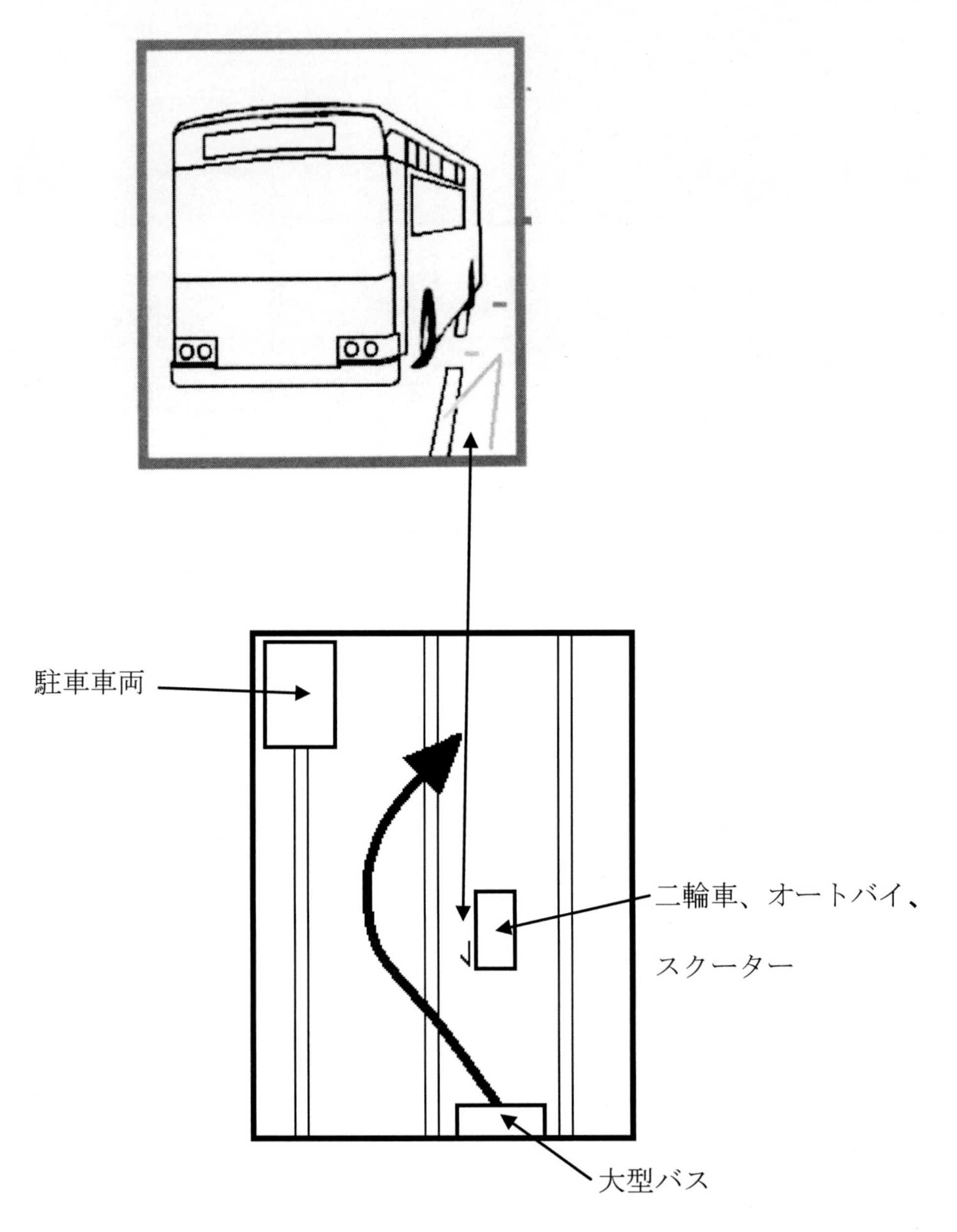

二輪車の左後方から大型バスが進路変更した時に前方の駐車車両を避ける為に右に進路変更する。この時にバックモニターの図形ラインで接触を予知。

⑵　右側からの接触

交差点の手前を走行中に右側から右折車両との接触･巻き込み事故防止の確認

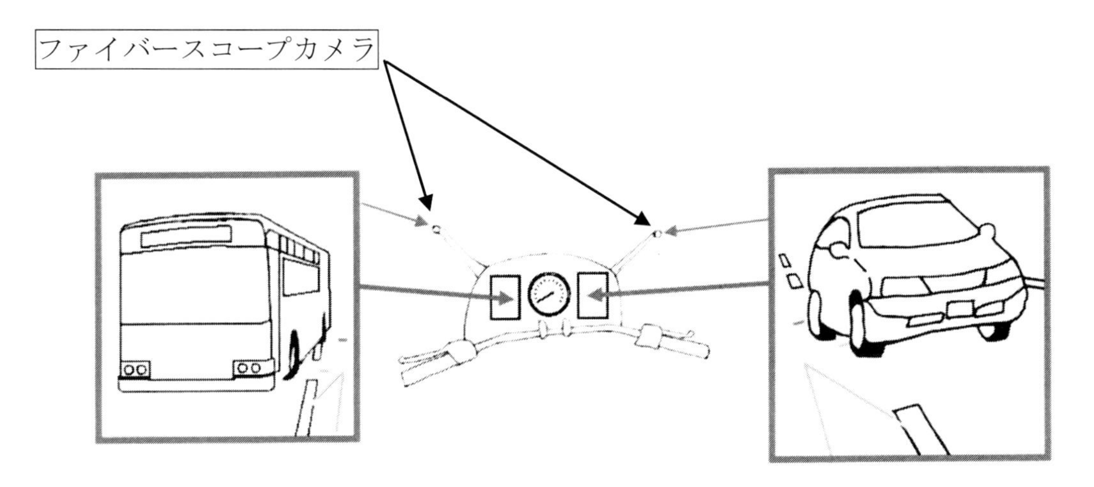

スピードメーターの左右にバックモニター、右後方を示す。

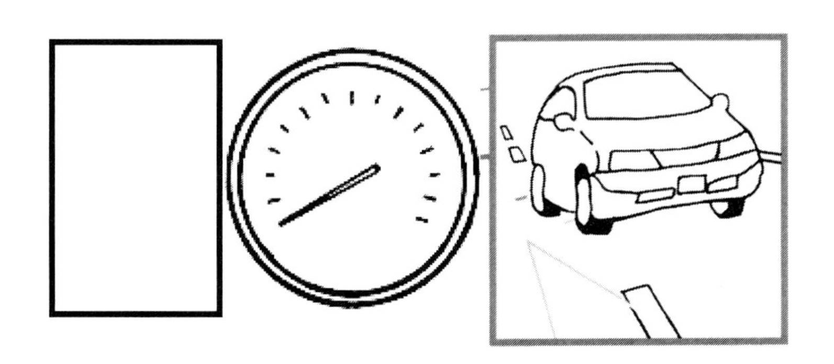

スピードメーターとタコメーターの左右にモニター、右後方を示す。

デジタルスピードメーターの左右にバックモニター、右後方を示す。

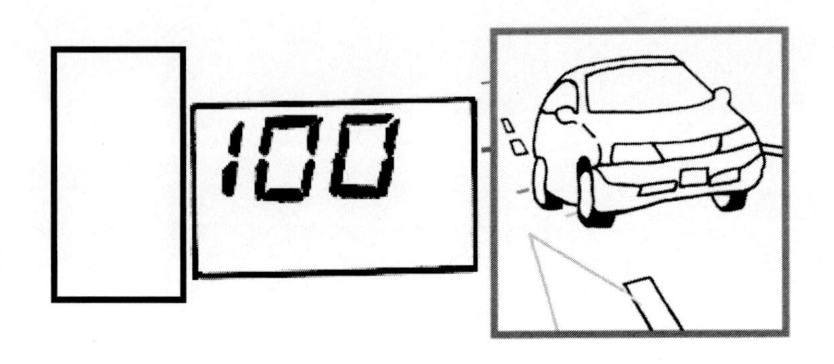

メーターパネル内の左右にバックモニター、右後方を示す。

交差点の左折事故防止

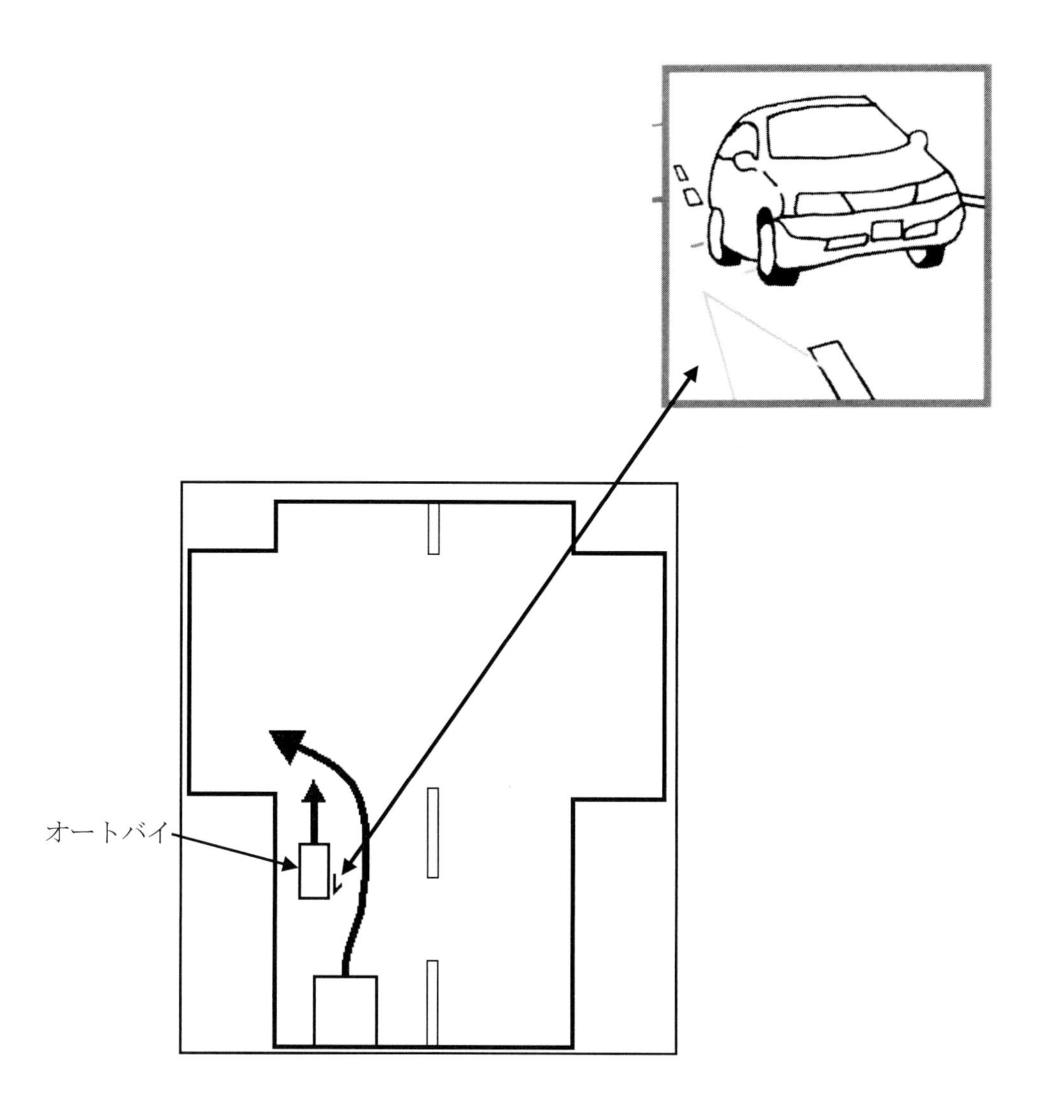

交差点の手前を二輪車、オートバイまたはスクーターが走行中に右後方からの左折車両による接触・巻き込み事故を避けるために右バックモニターの図形ラインの確認で予知。

第二章　電動車椅子

１，ファイバースコープバックモニター図形ラインの解説

電動車椅子のファイバースコープカメラはハンドルに取り付けられていた所に設置され、計器左右のバックモニターで障害物、段差などが確認でき見られる。

２，歩道の電柱と塀の間を通り抜け

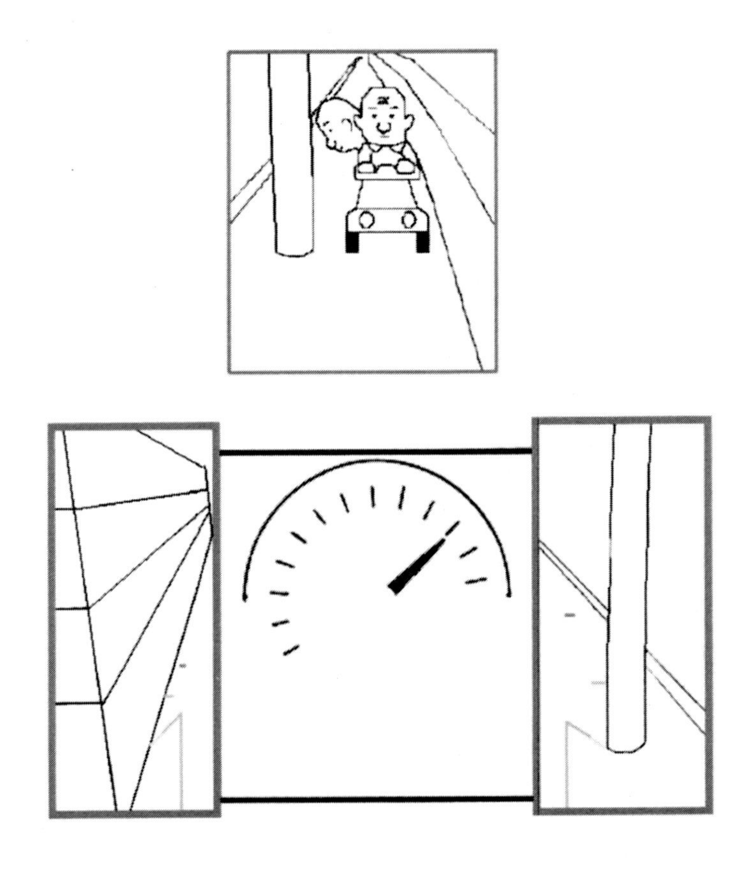

通り抜ける左右の間を確認する場合、計器左右のバックモニターなら図形ラインで容易に確認できる。

３，転倒・転落事故

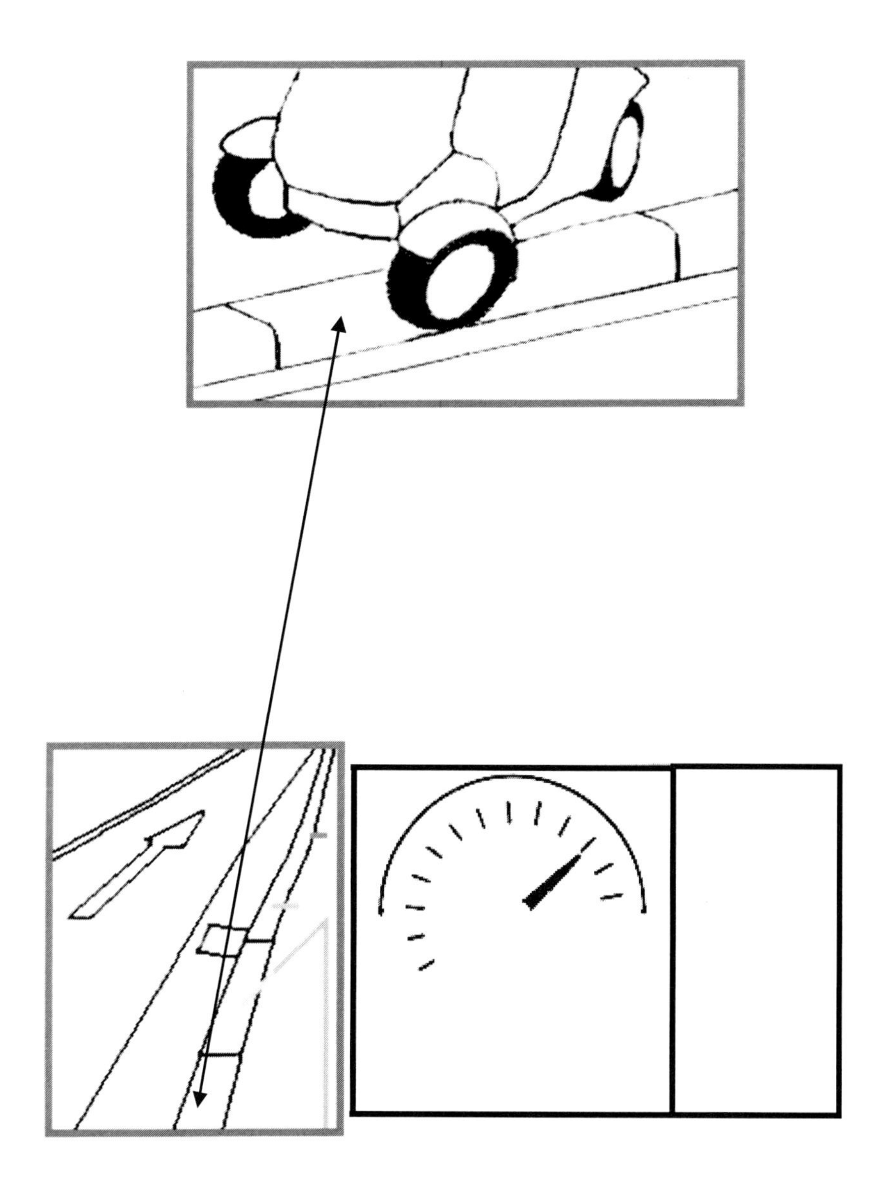

歩道の段差により、側溝などに転倒する事故、用水路、水田に転落する事故がありますが､バックモニターの図形ラインで容易に安全確認ができるから事故防止ができる。

4，歩道の障害物・電柱

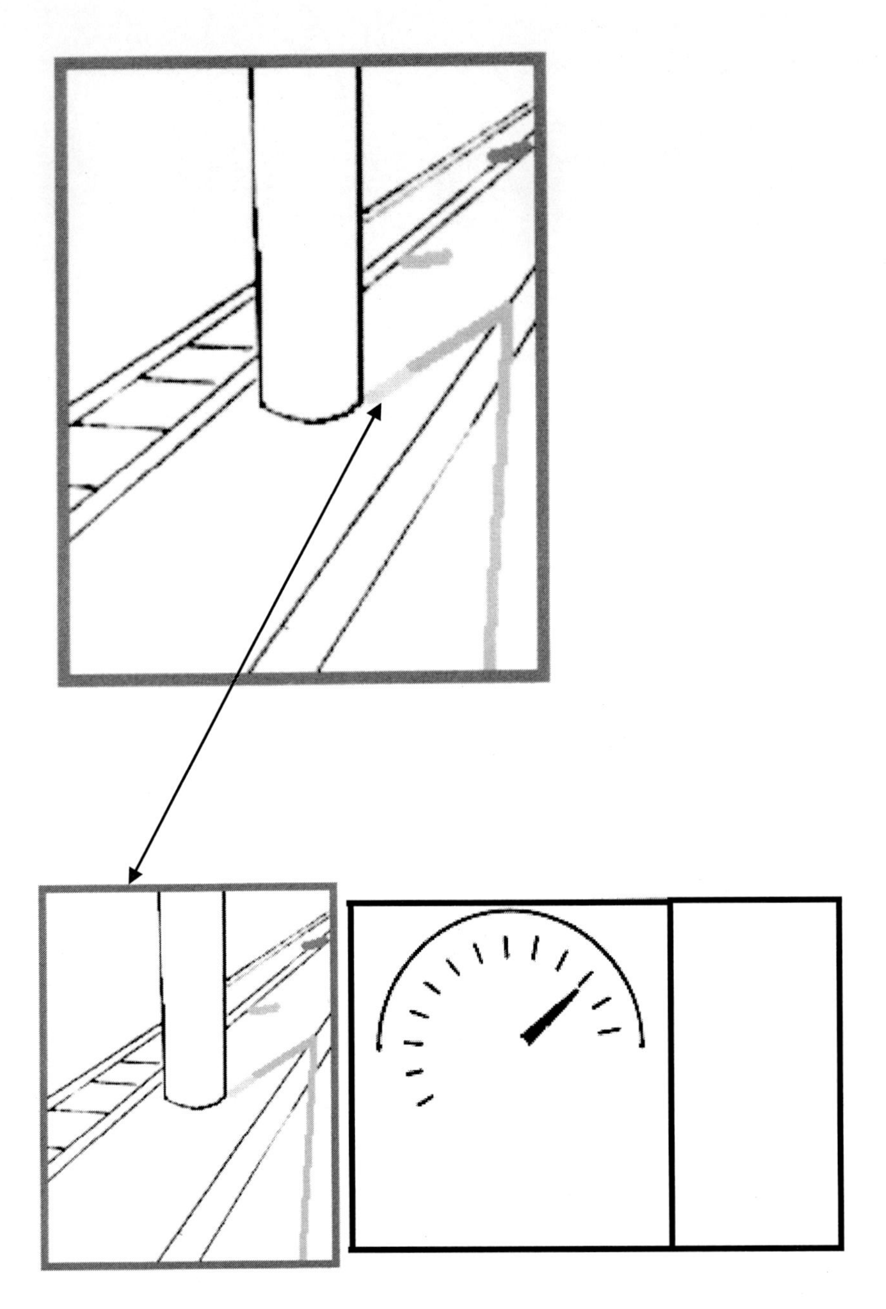

歩道の障害物、電柱を避けて、安全に走行する場合もバックモニターの図形ライン

で間を確認できるから接触の心配がなく、車道に出ることも押さえられる。

５，横断歩道の転倒防止

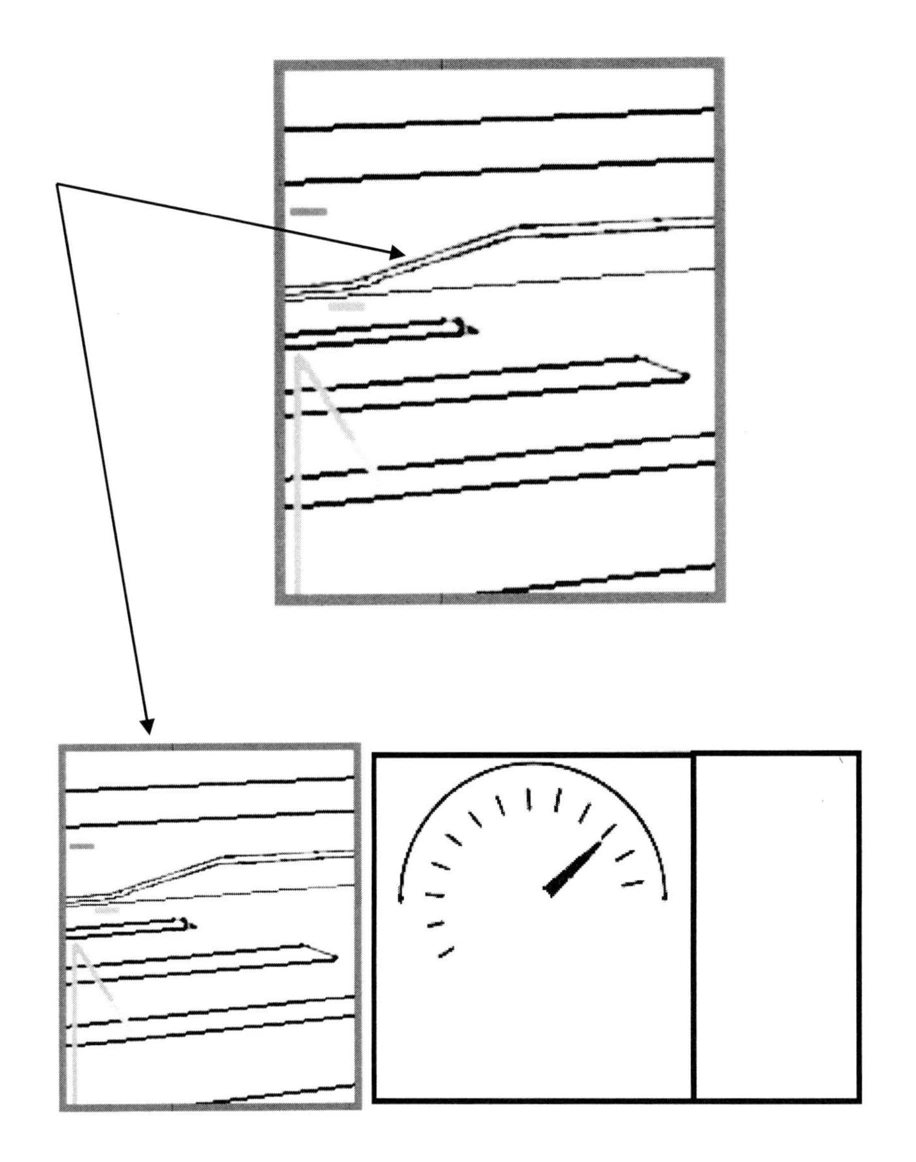

横断歩道の曲がり角による段差をバックモニター図形ラインで確認できるので転倒防止を避けられる。

English description

Chapter 1 Two-wheeled Vehicle, Motorcycle, Scooter

1

Description of a fiberscope back monitor figure line

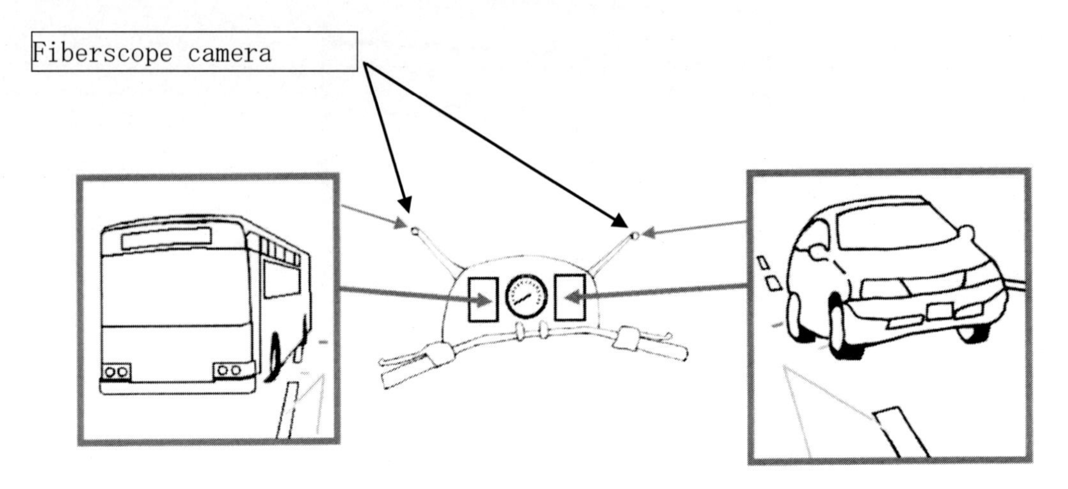

A two-wheeled vehicle, a motorcycle, and the fiberscope camera of a scooter are installed in the place attached in the handle, and are seen by the back monitor.

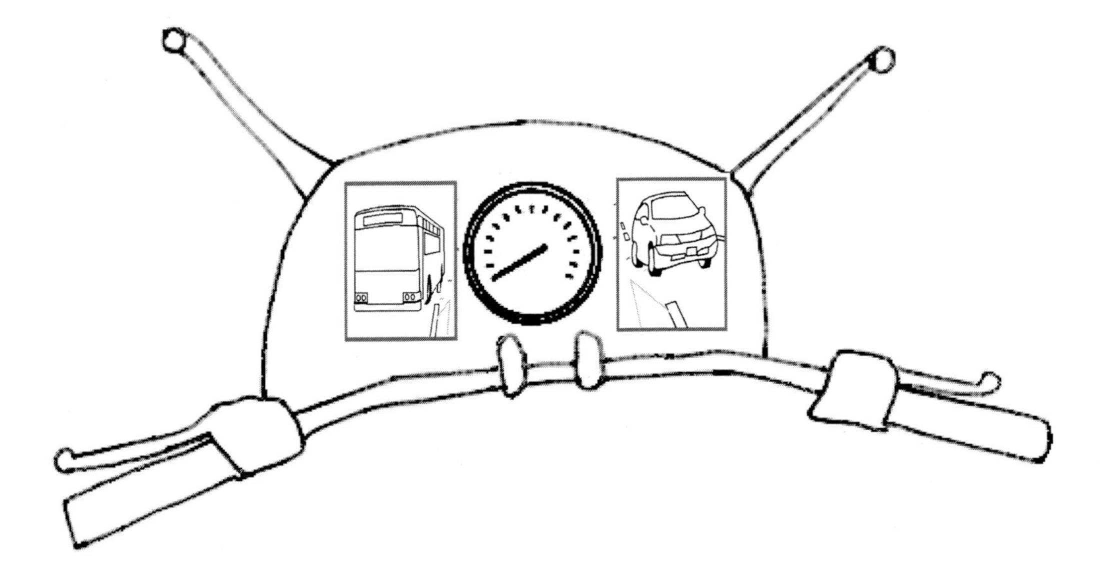

2

The pattern of a minor collision, passing, a crossing(illustration explanation)

(1)

Left-hand side contact

The check of the minor collision prevention by the vehicles which pass a motorcycle changing a course

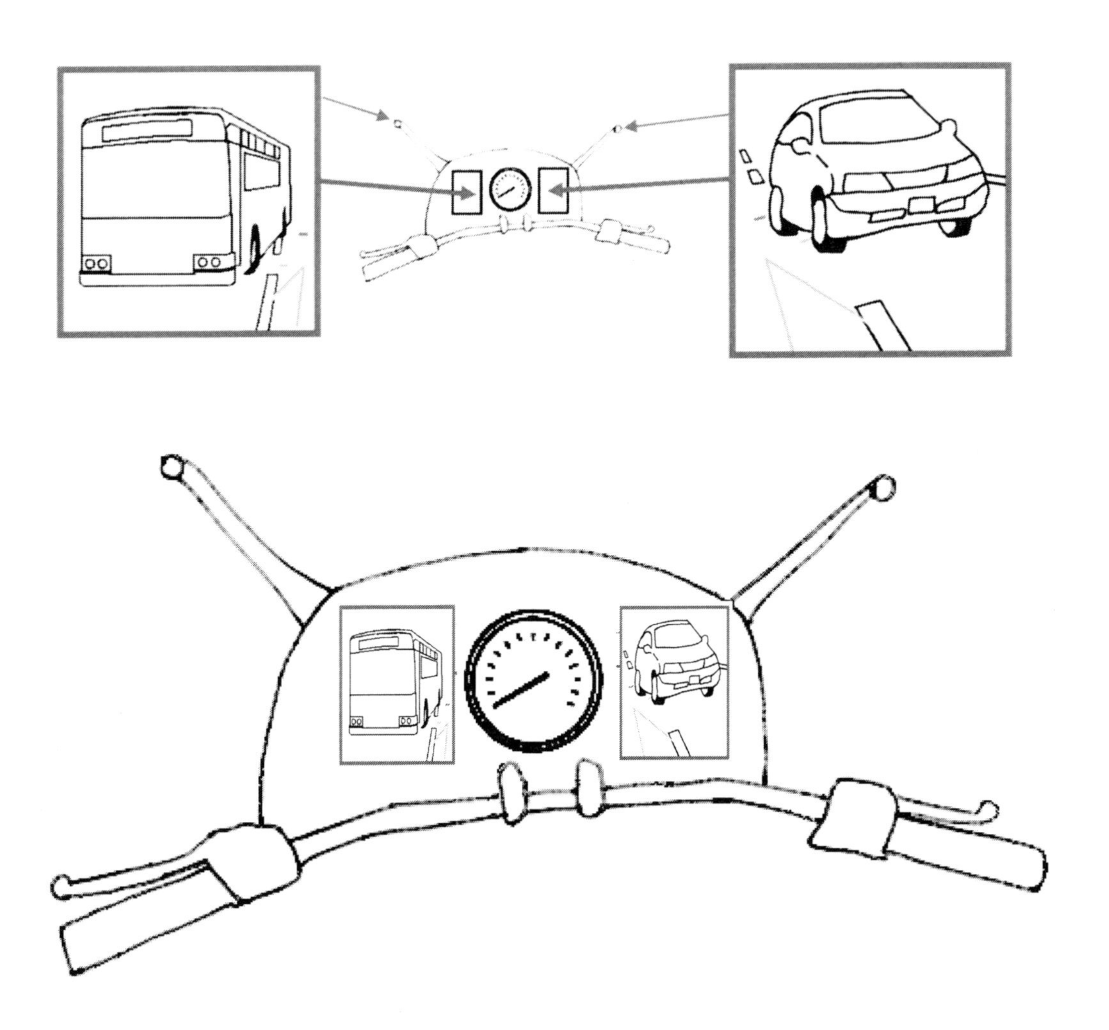

A pack monitor and the method of the left rear are shown in right and left of a speedometer.

A monitor and the method of the left rear are shown in right and left of a speedometer and a tachometer.

A back monitor and the method of the left rear are shown in right and left of a digital speedometer.

A back monitor on either side and the back are shown in a meter panel.

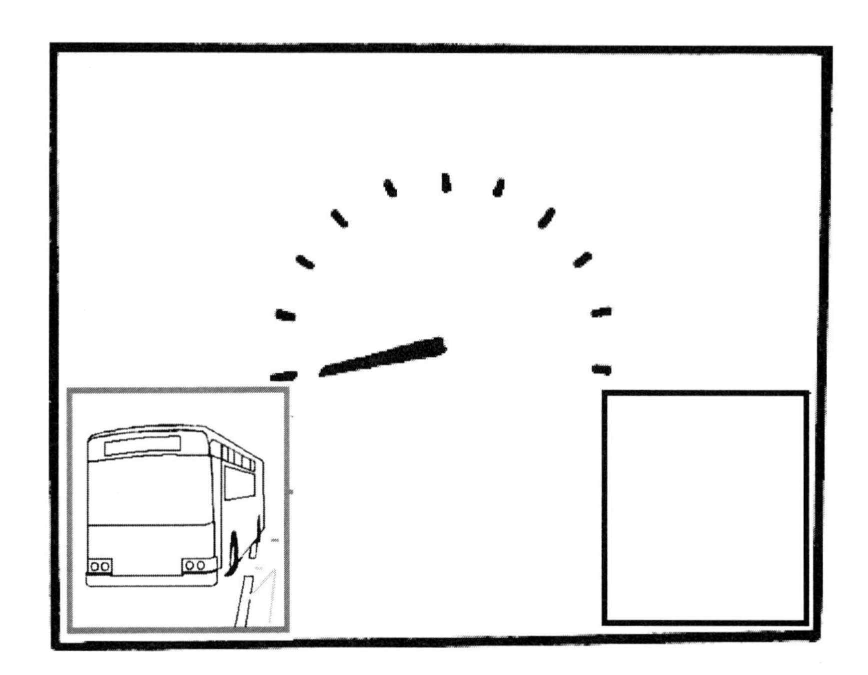

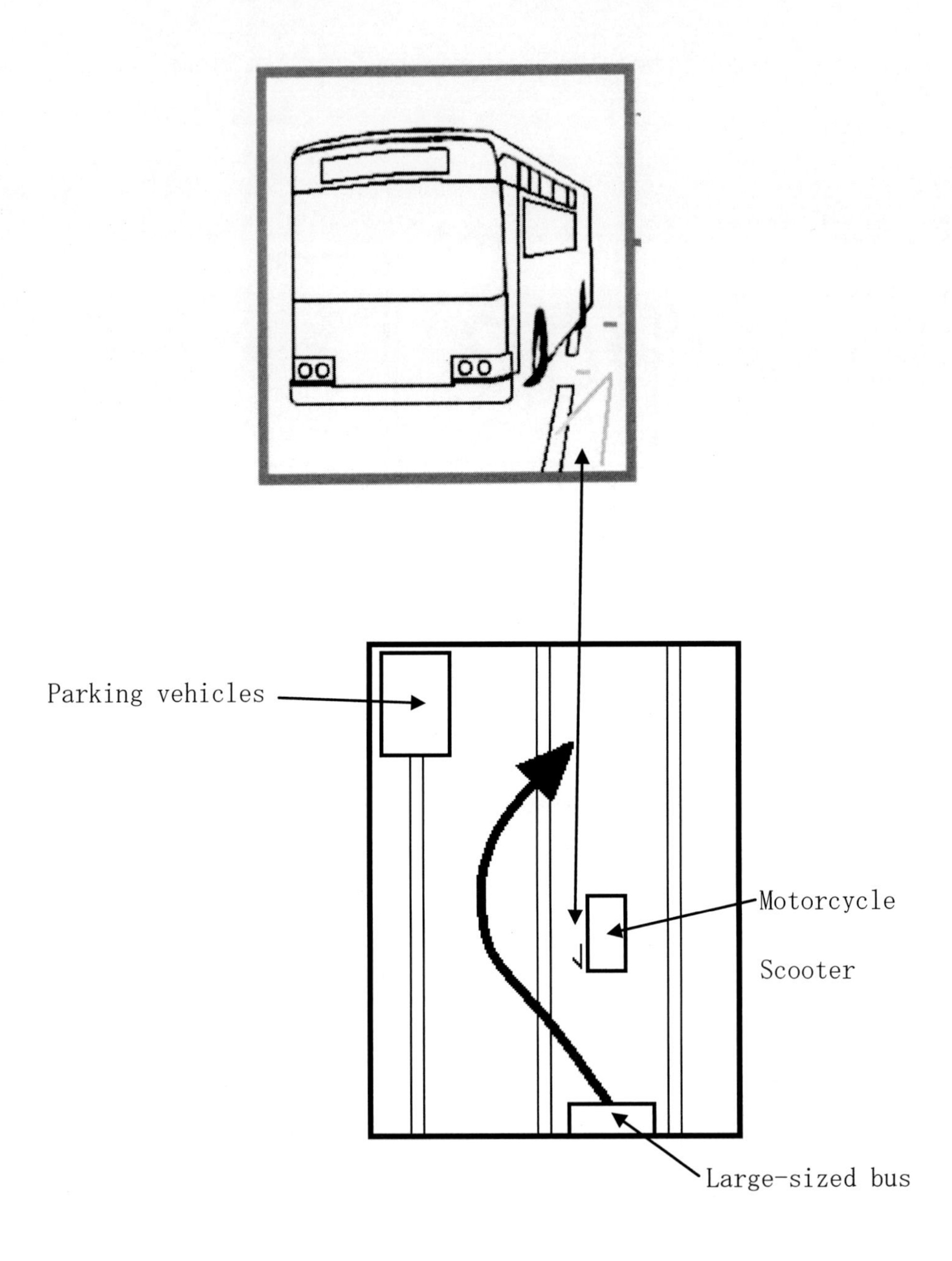

When a large-sized bus makes a course change from the method of the left rear of a two-wheeled vehicle, in order to avoid front parking vehicles, a course change is made on the right.

At this time, contact is foreknown with a back monitor's figure line.

(2)

Right-hand side contact

While running intersectional this side, it is the check of contact and involvement accident prevention with right-hand side to right-turn vehicles.

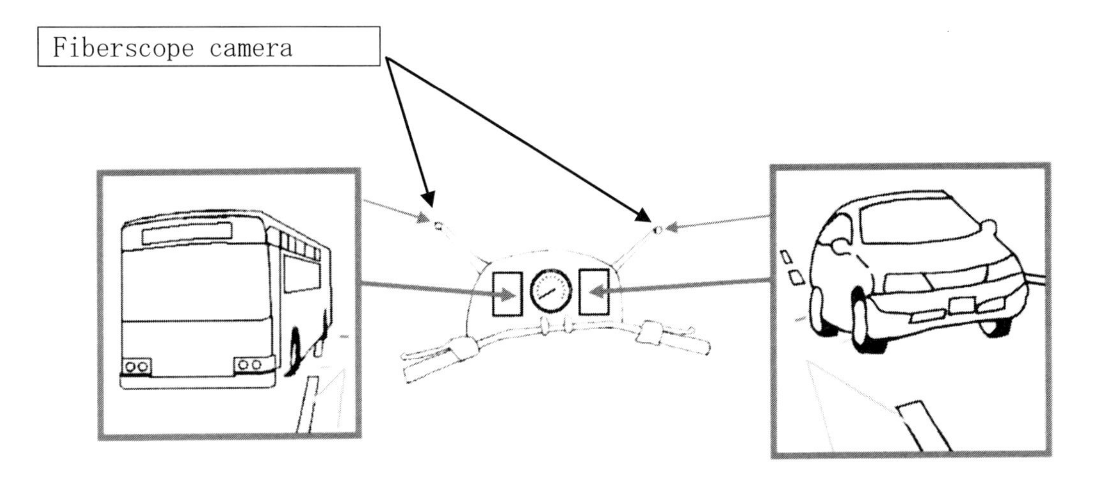

A pack monitor and the right back are shown in right and left of a speedometer.

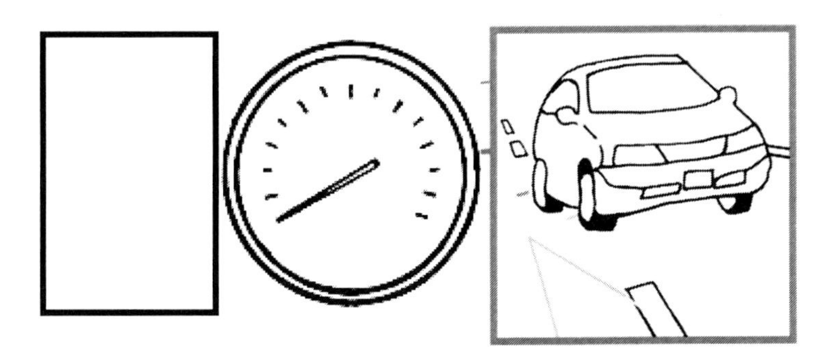

A monitor and the right back are shown in right and left of a speedometer and a tachometer.

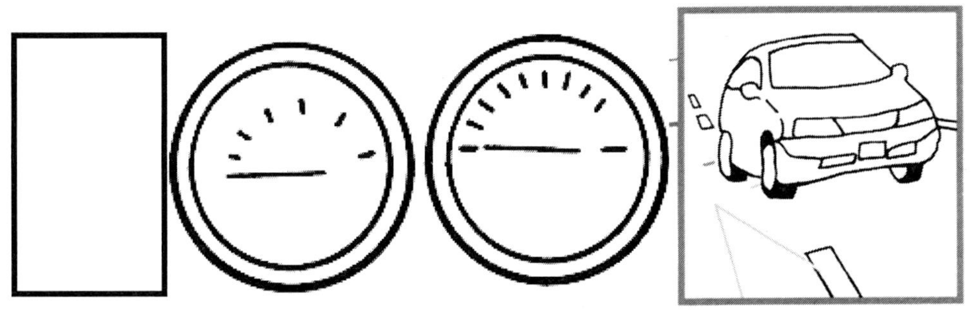

A back monitor and the right back are shown in right and left of a digital speedometer.

A back monitor and the right back are shown in the right and left in a meter panel.

Left turn of intersection, accident prevention

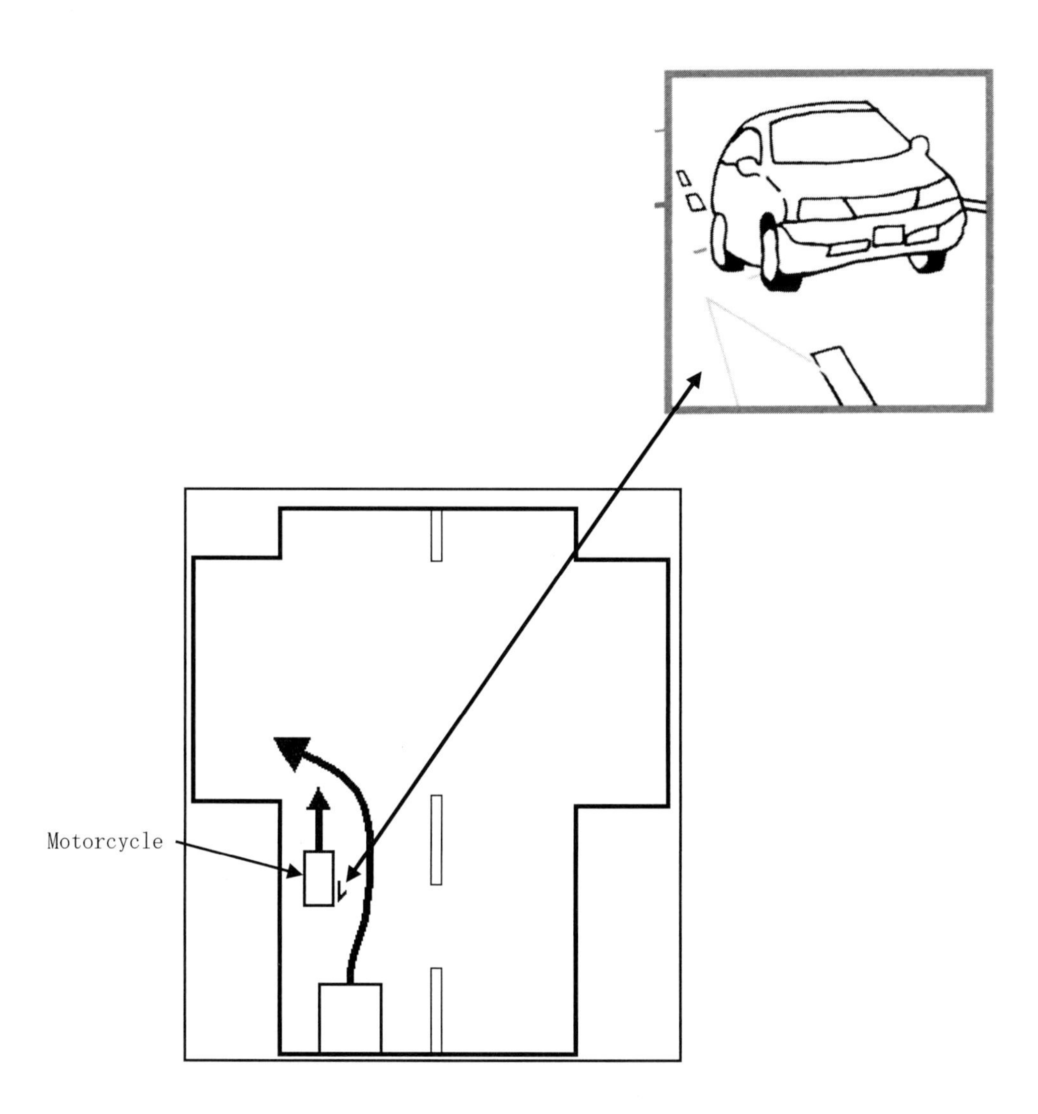

While a two-wheeled vehicle, a motorcycle, or a scooter runs intersectional this side, in order to avoid the contact and the involvement accident by the left-turn vehicles from the right back, it foreknows by the check of a right back monitor's figure line.

Chapter 2 Electric Wheelchair

1,

Description of a fiberscope back monitor figure line

The fiberscope camera of an electric wheelchair is installed in the place attached in the handle, and it can be tried by the back monitor of meter right and left to check an obstacle, a level difference, etc.

2,

It passes between the telegraph pole of a sidewalk, and a fence.

When passing, between right and left can be checked with a back monitor's figure line.

3,

Accident of a fall and falling down

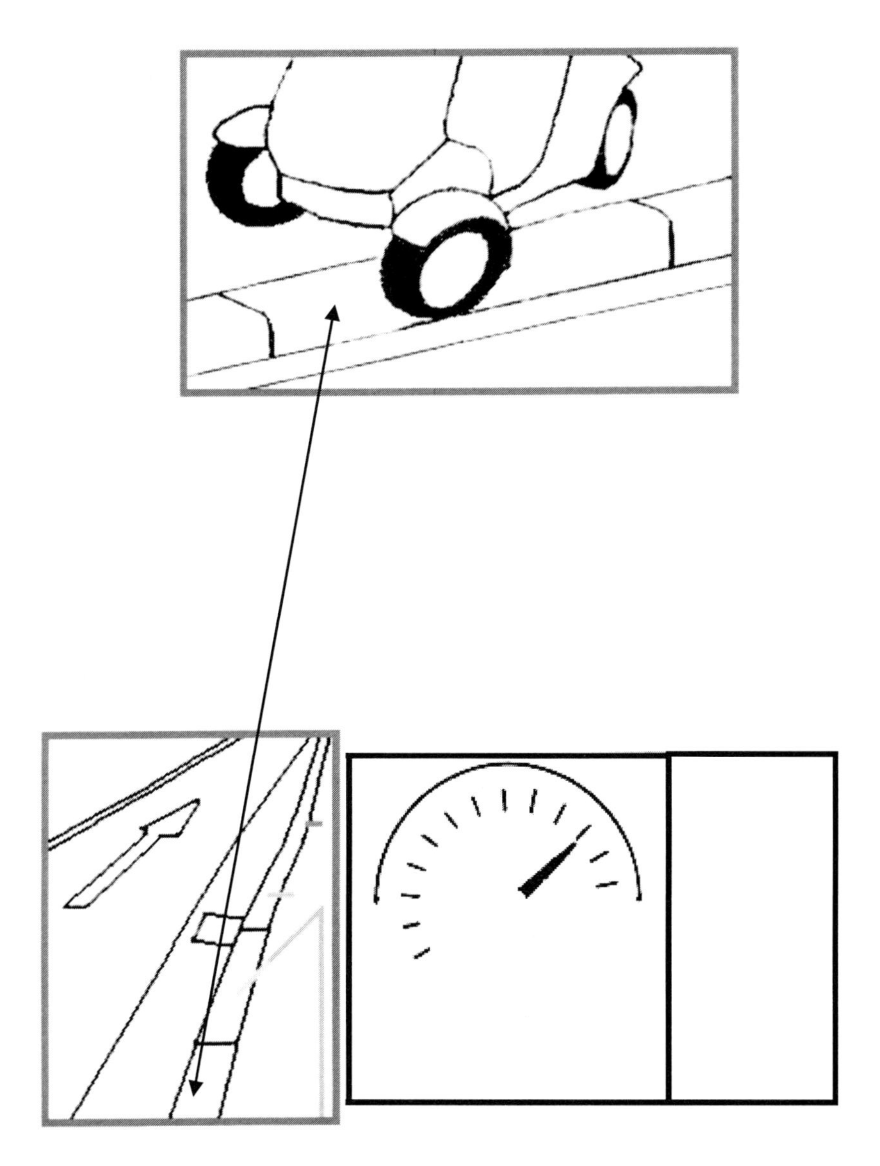

Although there is an accident which falls in the accident reversed to a sewer etc., an irrigation canal, and a paddy field with the level difference of a sidewalk, since a safe check can be easily performed with a back monitor's figure line, accident prevention can be performed.

4,

The obstacle and telegraph pole of a sidewal

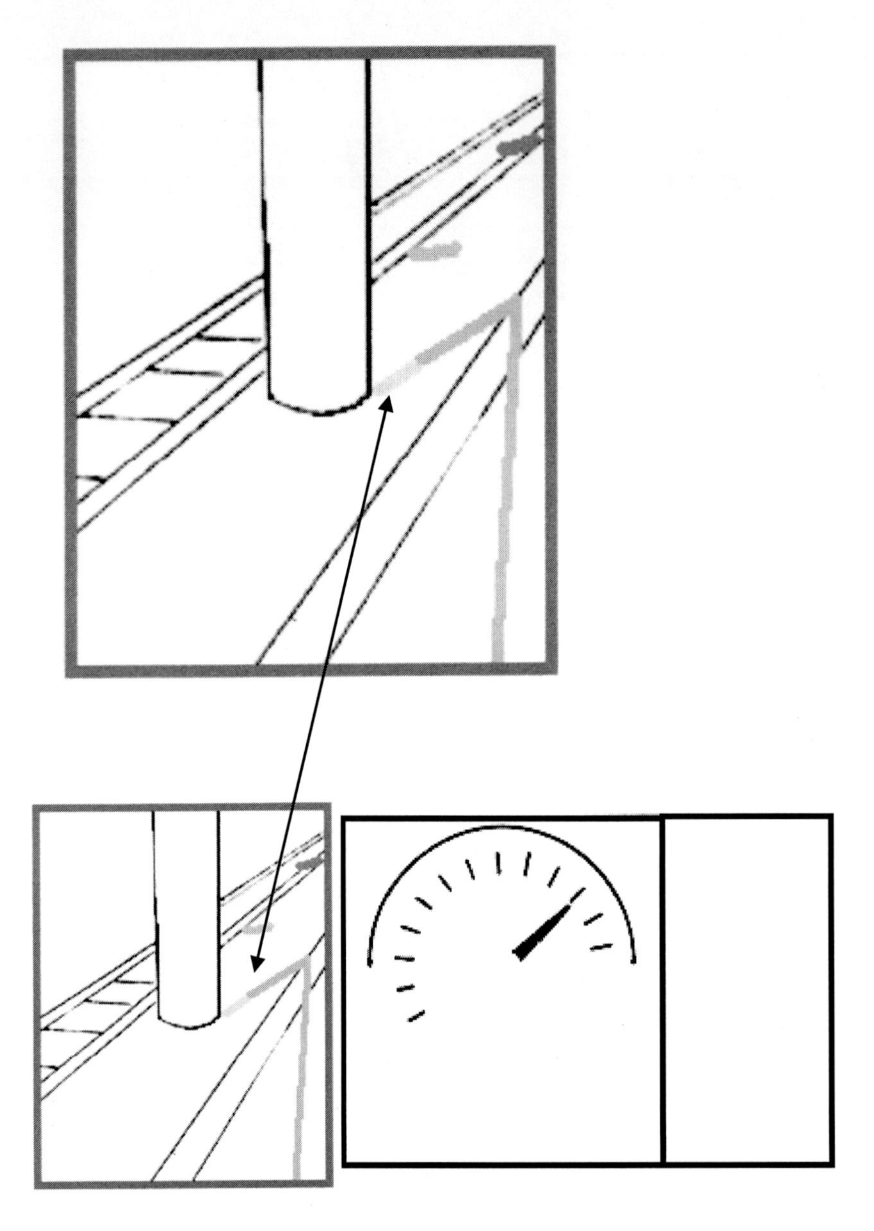

The obstacle of a sidewalk and a telegraph pole are avoided, since between can be checked with a back monitor's figure line when running safely, there are no worries about contact, and coming out to a driveway is also pressed down.

5,

It prevents breaking down from a sidewalk.

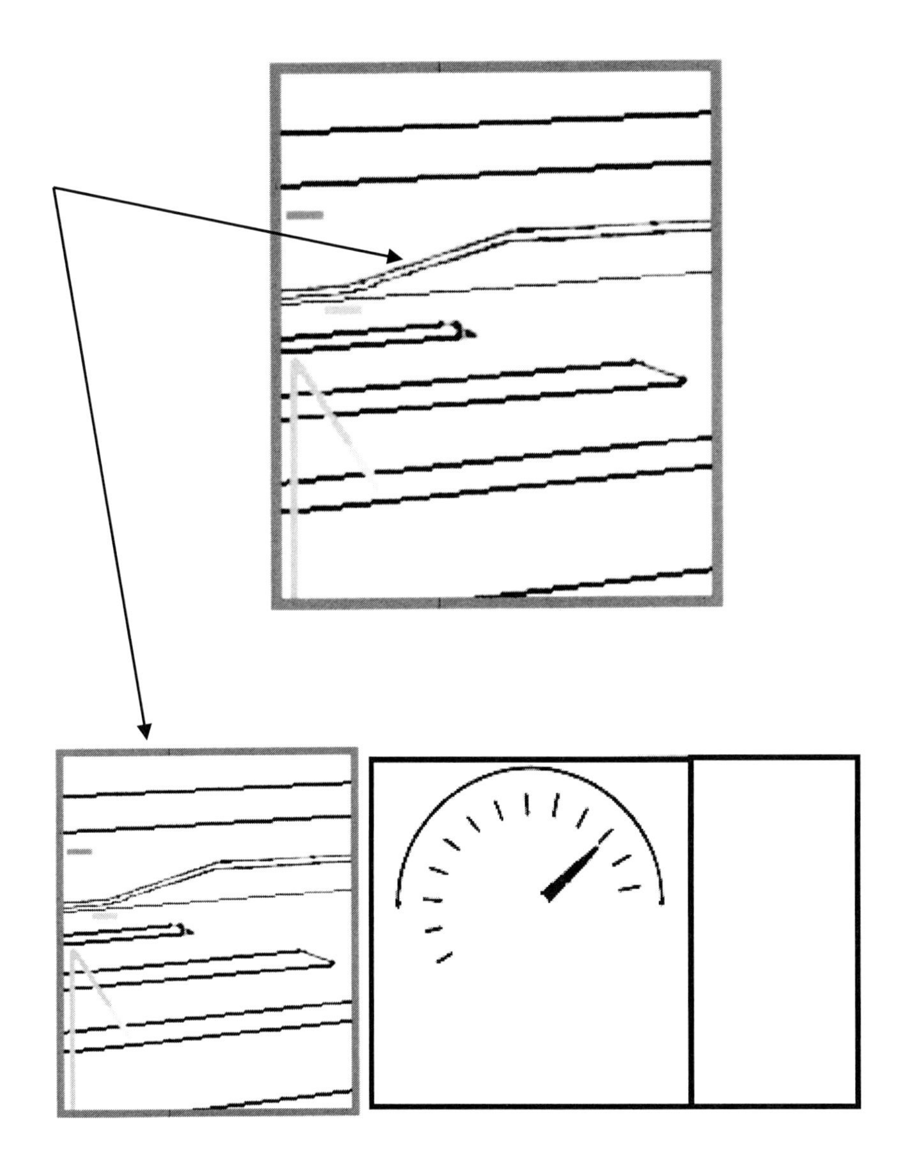

Since the level difference by the corner of a street of a pedestrian crossing can be checked with a back monitor figure line, fall prevention is avoidable.

中文解说

第一章摩托车以及低座轻型摩托车

１，

纤维镜背监视器图形线的解说

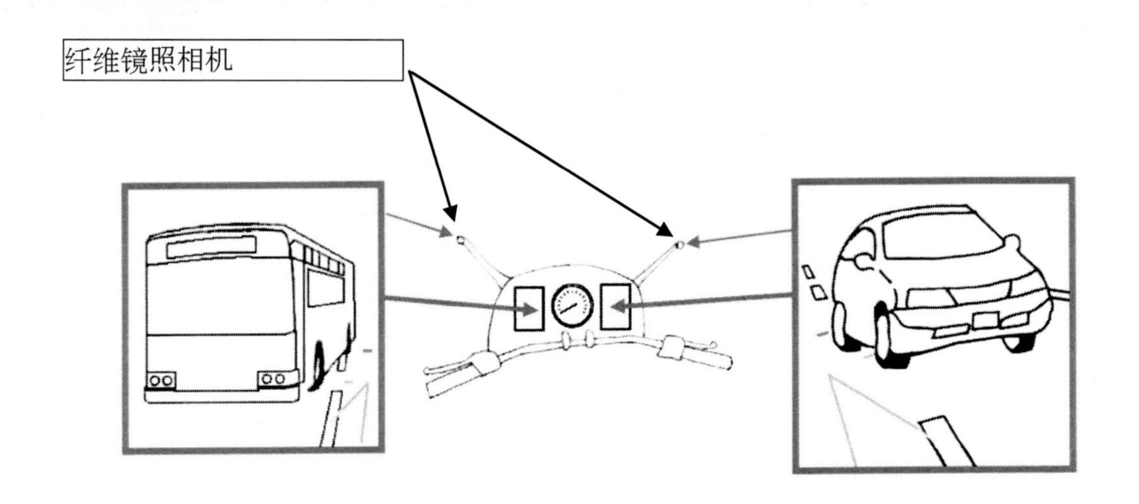

双轮车，摩托车，低座轻型摩托车的纤维镜照相机被在被对方向盘安装的地方设置，用仪器左右的背监视器可以看到。

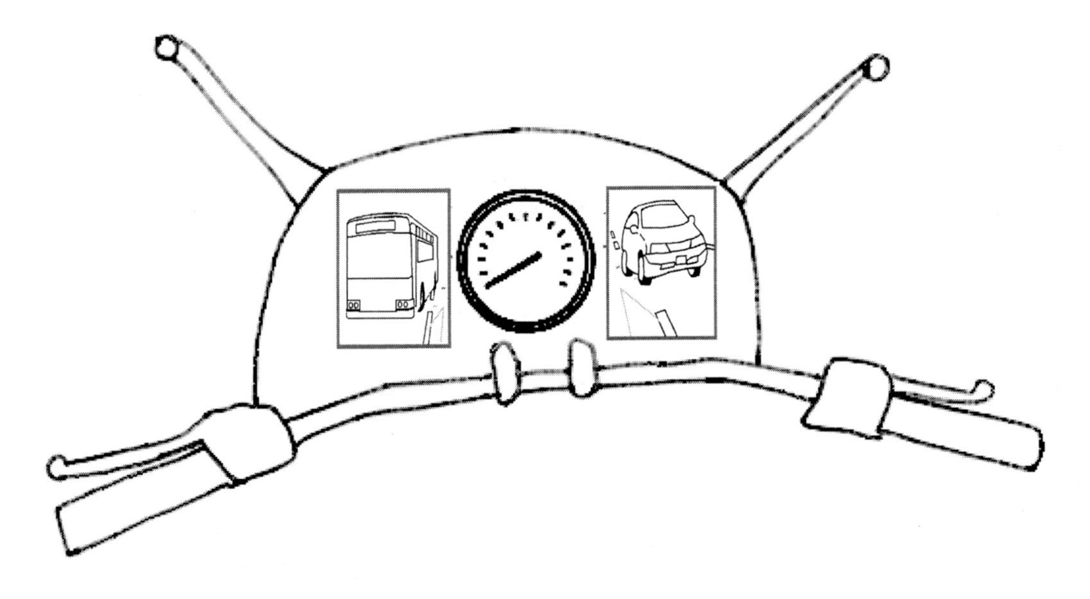

２、

接触事故的模式、超过、左转弯、右转弯、十字路口(插图解说)

⑴

左侧的接触

出自超过摩托车的车辆改变前进的道路的的接触事故防止的确认

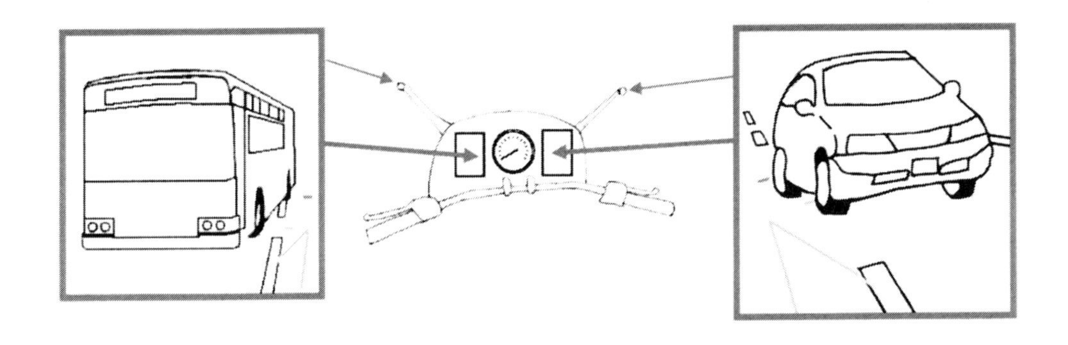

对速度计的左右显示包监视器，左后方。

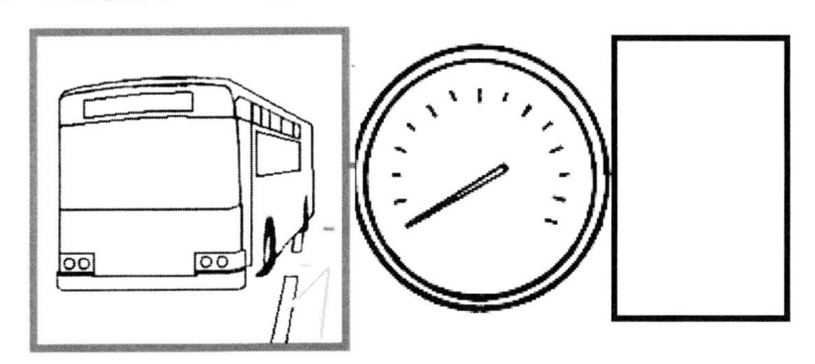

对速度计和旋速计的左右显示监视器，左后方。

对数码的速度计的左右显示背监视器，左后方。

对测量仪器面板里面的左右显示背监视器，左后方。

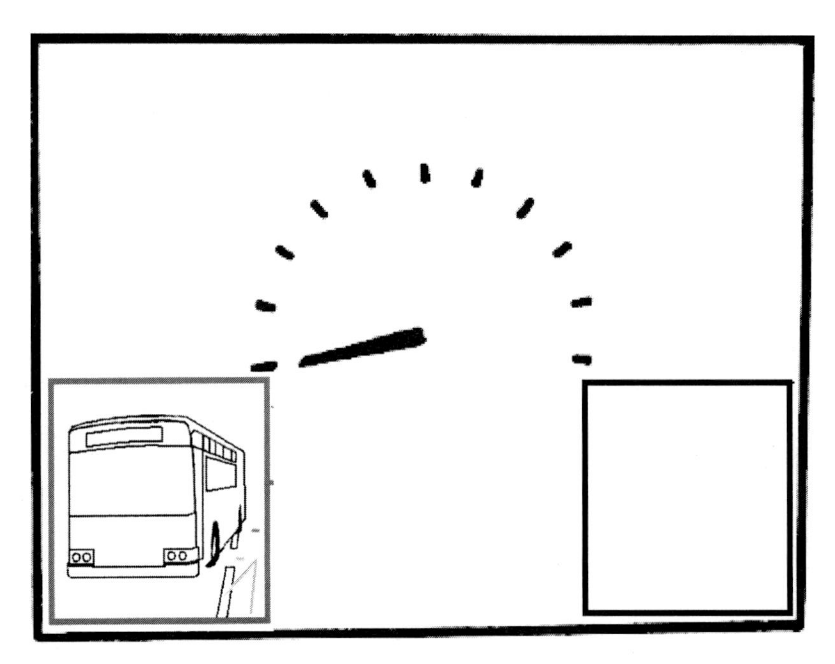

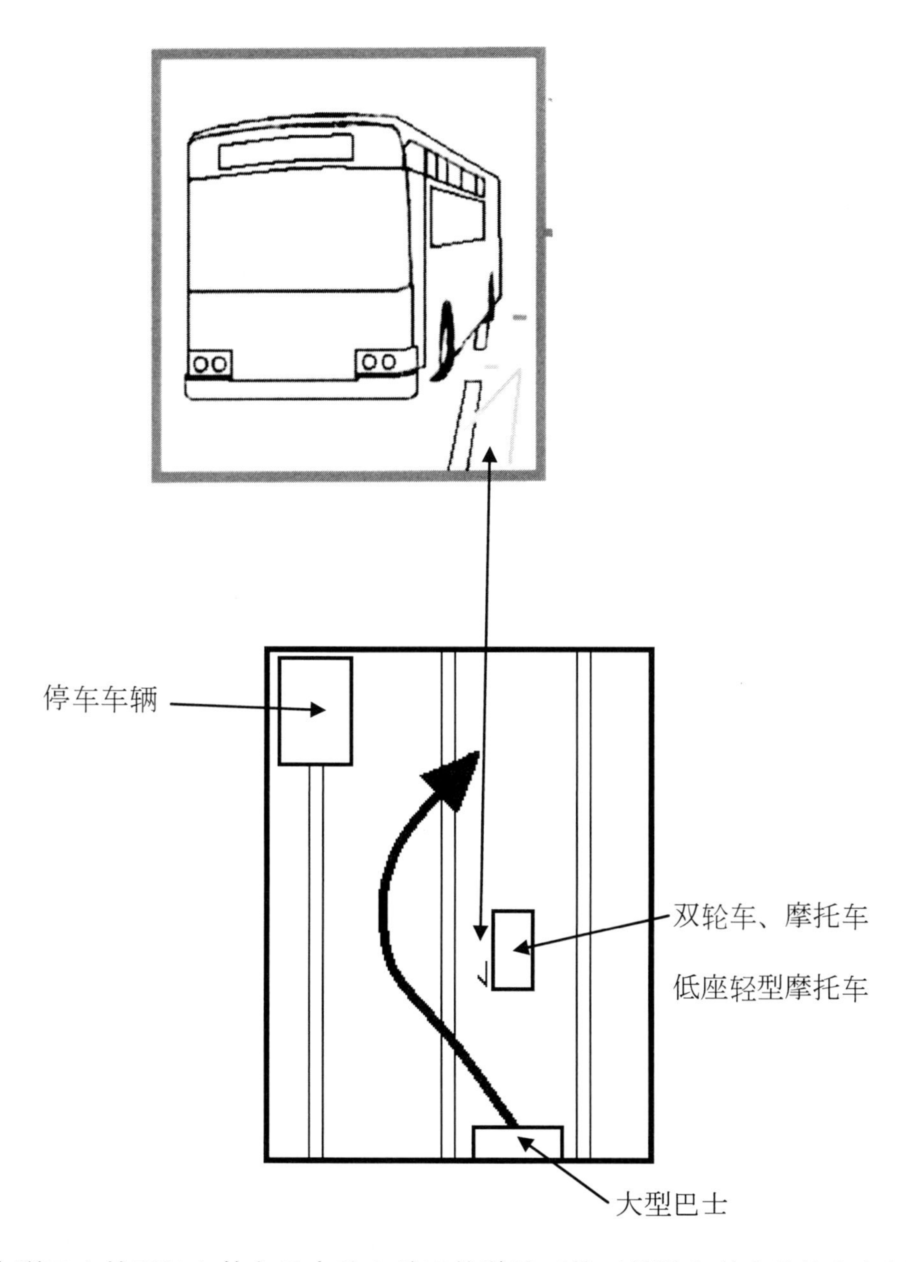

为在大型巴士从双轮车的左后方改变前进的道路了的时候避免前方的停车车辆向右改变前进的道路。 在这时候，用背监视器的图形线预见接触。

⑵

右侧的接触

被在行驶时在快到十字路口的地方从右侧确认与左转弯车辆的接触、内卷事故防止

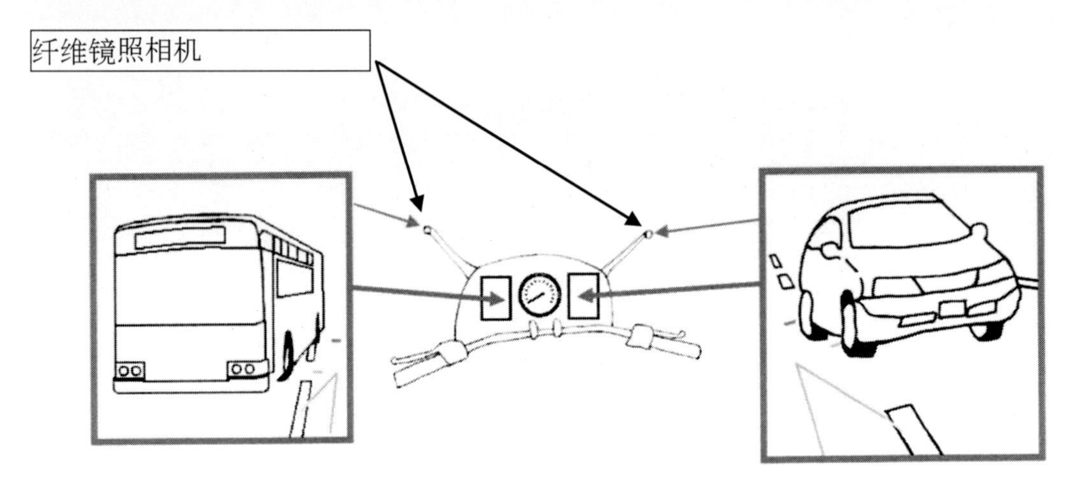

对速度计的左右显示包监视器，右后方。

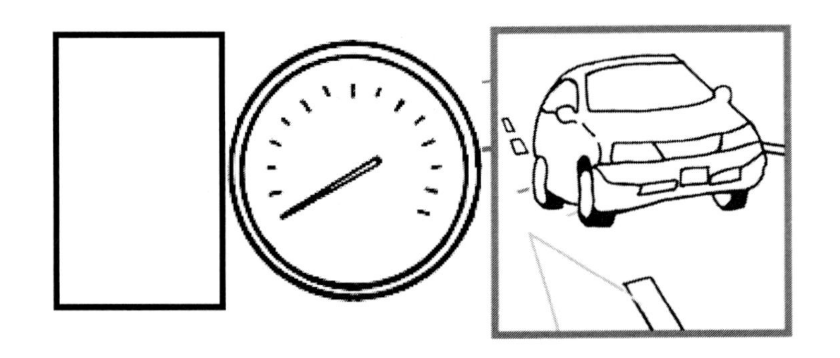

对速度计和旋速计的左右显示监视器，右后方。

对数码的速度计的左右显示背监视器，右后方。

对测量仪器面板里面的左右显示背监视器，右后方。

十字路口的左转弯事故防止

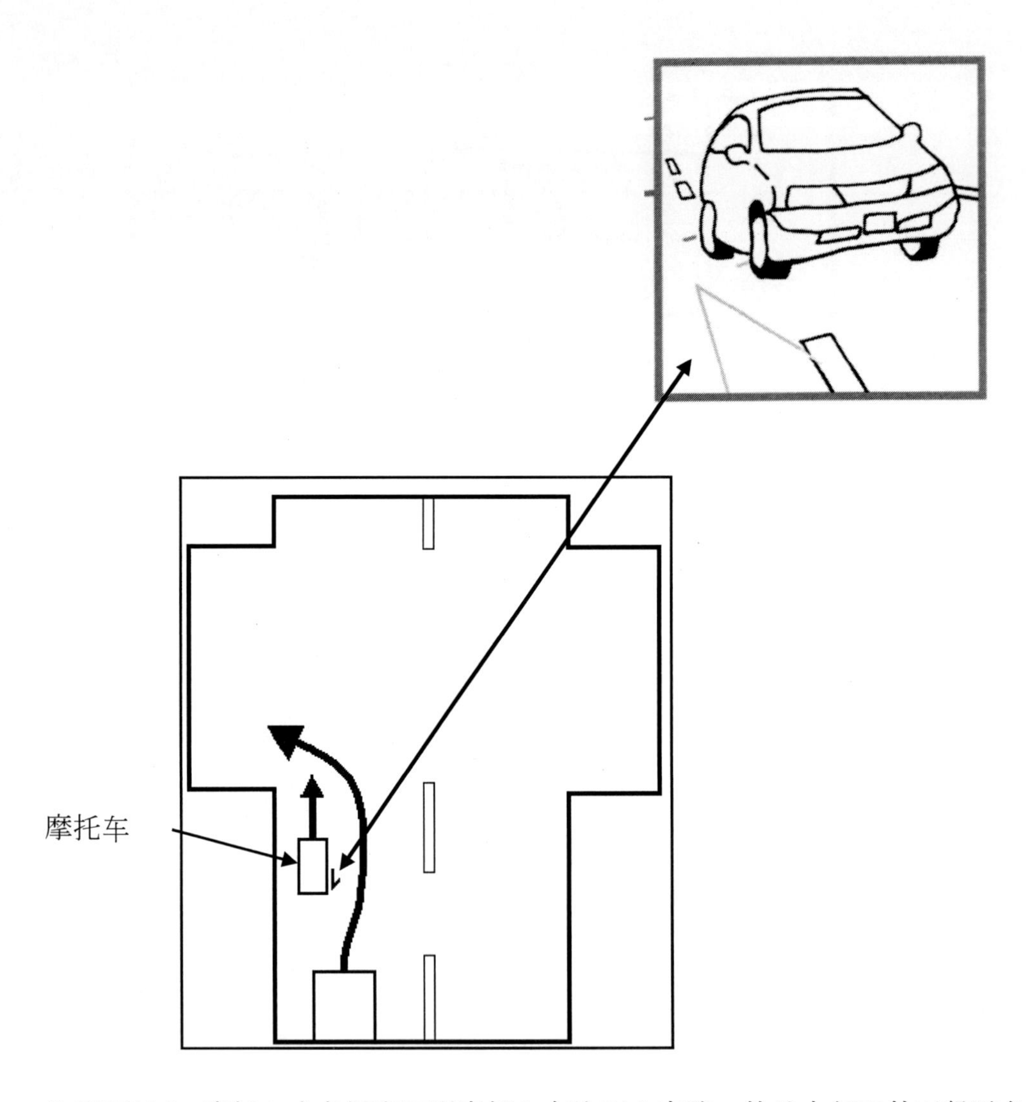

为当双轮车，摩托车或者低座轻型摩托车在快到十字路口的地方行驶的时候避免出自来自右后方的左转弯车辆的接触、内卷事故是右背监视器的图形线的确认，并且预见。

第 2 章电动轮椅

１，

纤维镜背监视器图形线的解说

电动轮椅的纤维镜照相机被在被对方向盘安装的地方设置，能用仪器左右的背监视器

确认障碍物，高低差别，可以看到。

２，

在人行道的电线杆和屏之间穿过

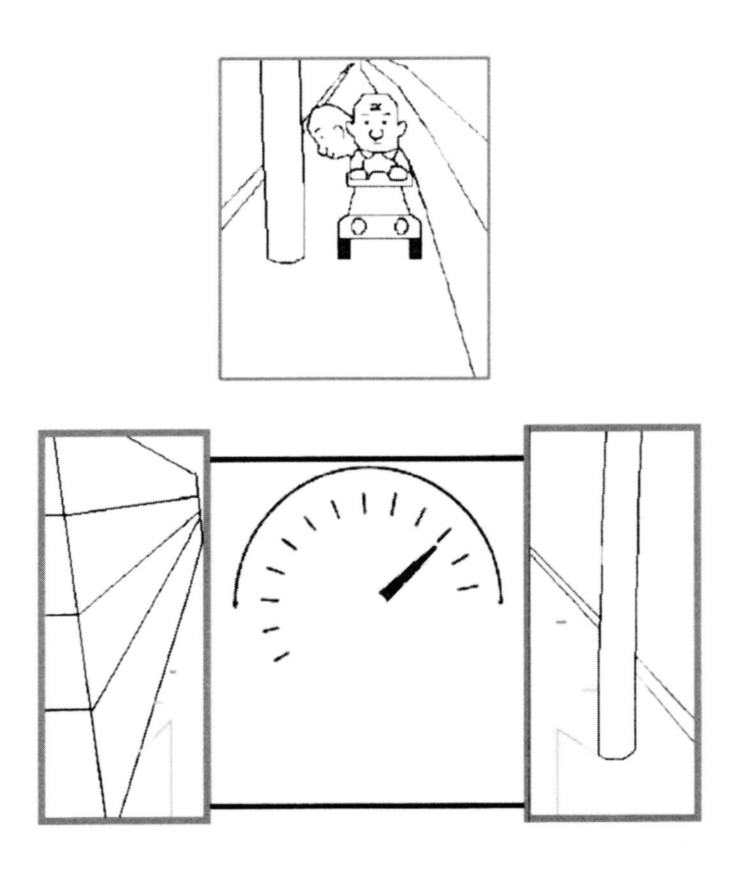

在在穿过的左右之间确认的时候，能用图形线容易确认仪器左右的背监视器。

３，

跌倒、坠落事故

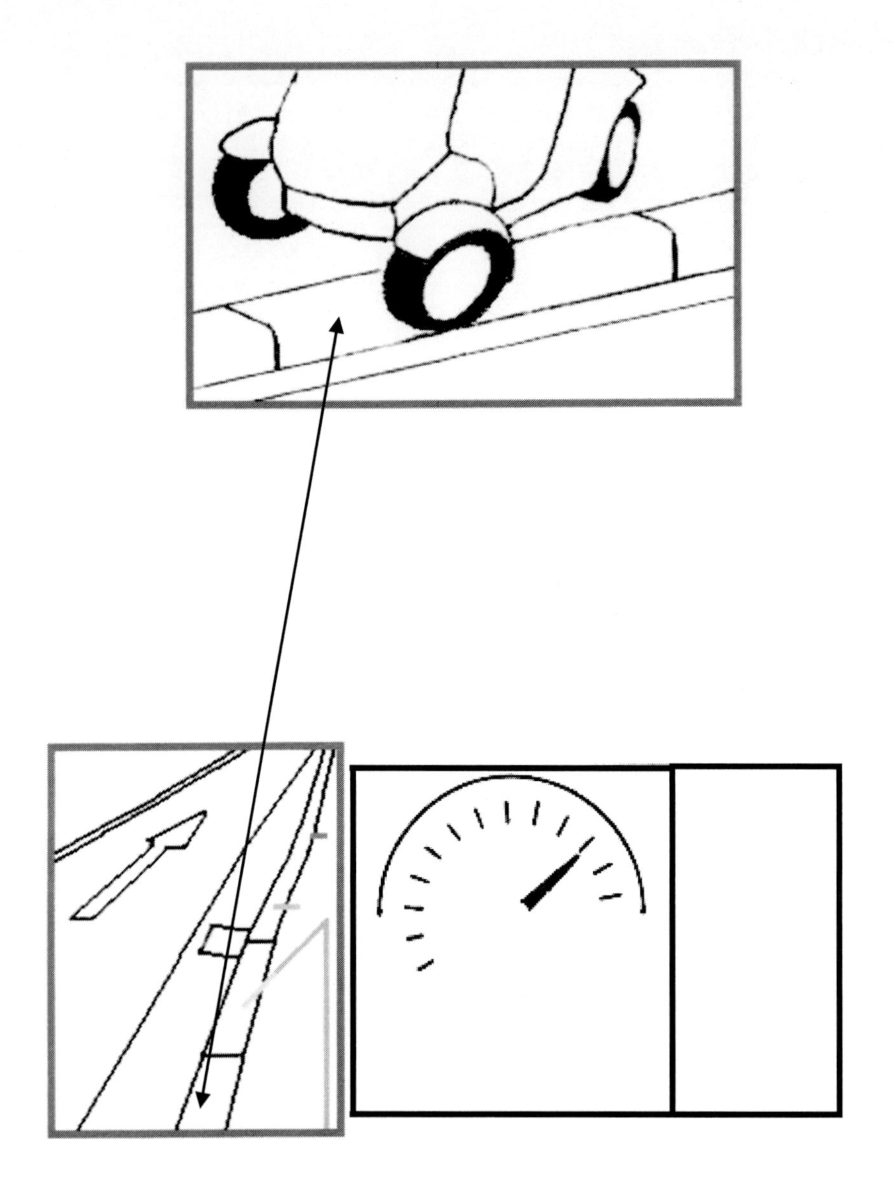

有坠落到在侧沟根据人行道的高低差别翻倒的事故，灌溉水渠，稻田的事故，但是因为进行安全确认容易能够是背监视器的图形线所以事故防止能够。

4，

人行道的障碍物、电线杆

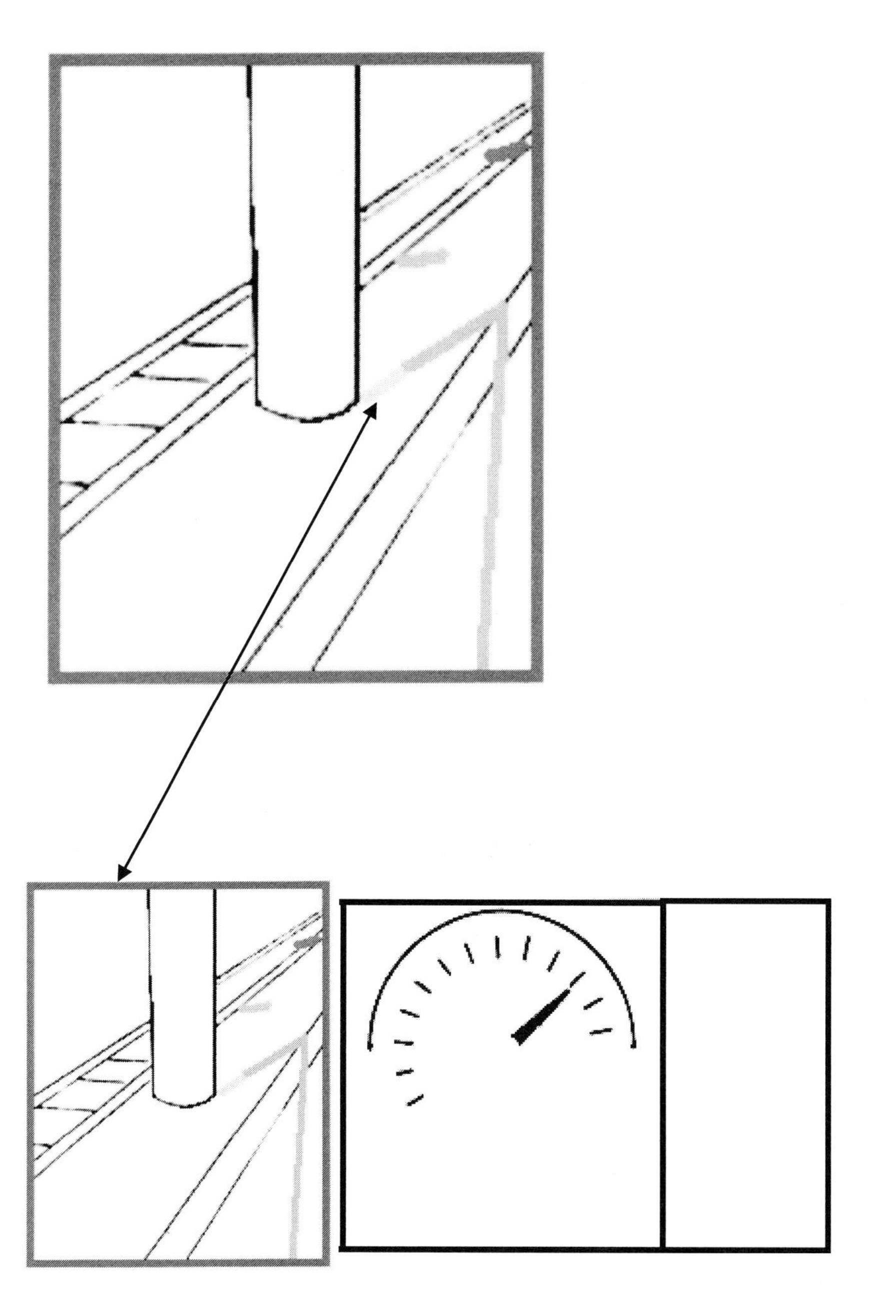

避免人行道的障碍物，电线杆，没有接触的担心，并且因为在安全地行驶的时候也能

在背监视器的图形线之间确认所以在车道出来的被控制。

5，

人行横道的跌倒防止

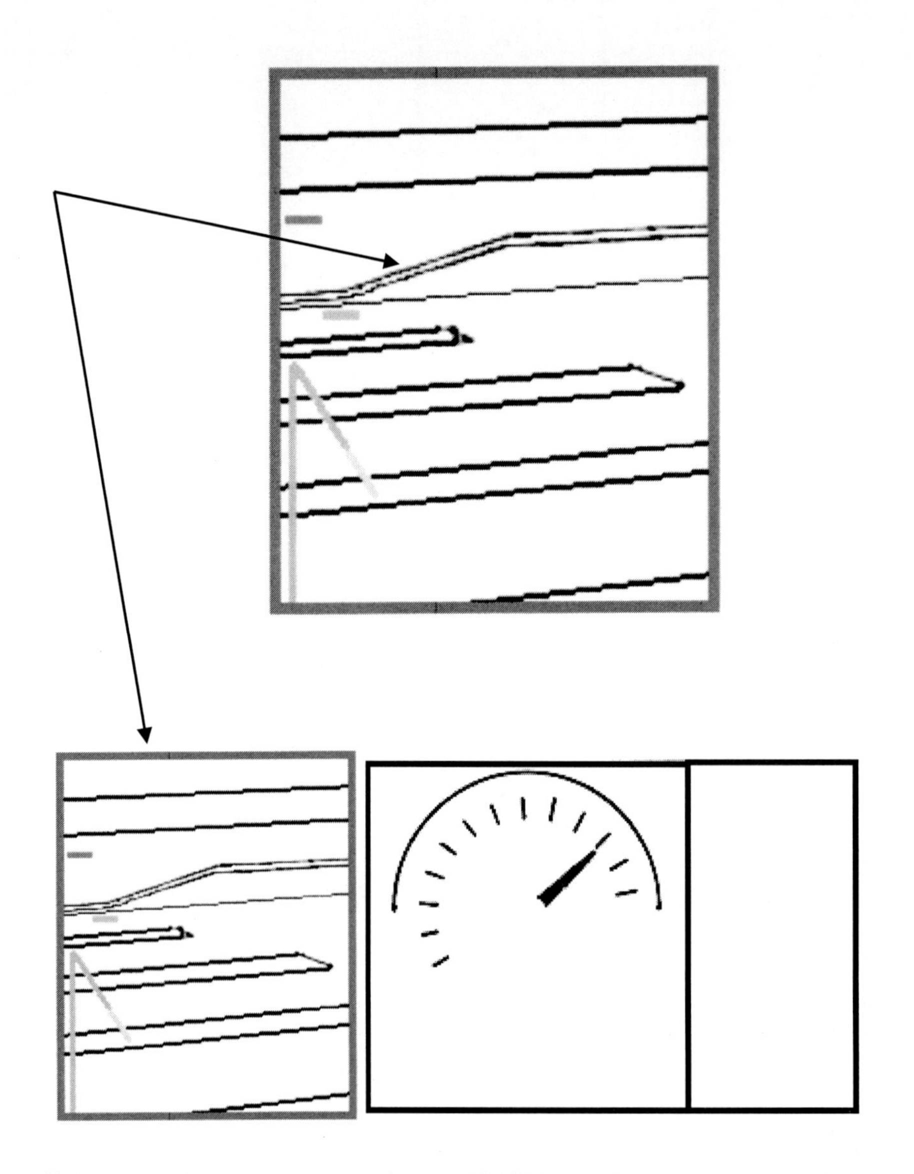

因为能用背监视器图形线确认出自人行横道的转角的高低差别所以被避免跌倒防止。

Deutscher Kommentar

Kapitel 1 zweirädriges Fahrzeug, Motorrad, Motorroller

1,

Der Kommentar des fiberscope kontrolliert zurück Zahl-Linie

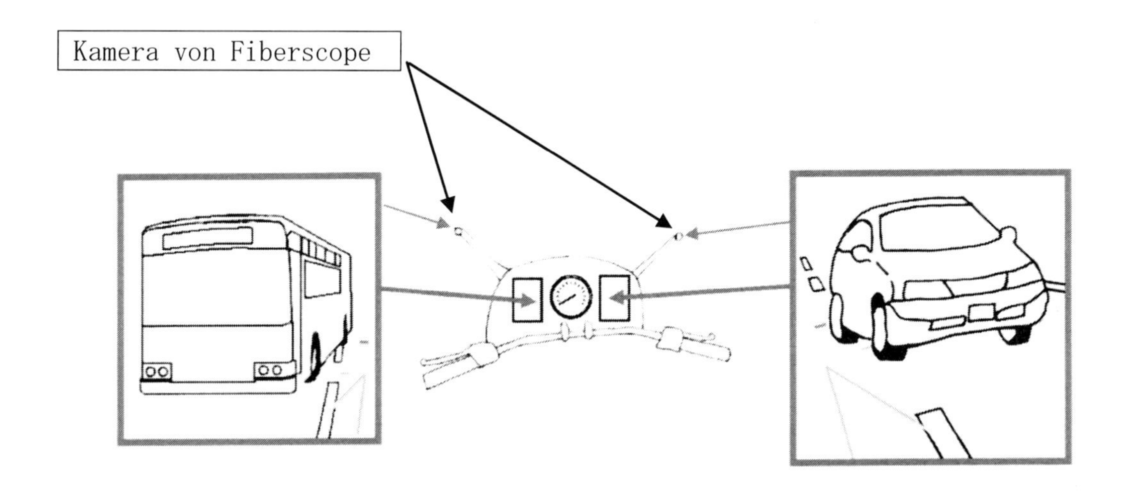

Der Platz, wo die fiberscope Kamera des zweirädrigen Fahrzeugs der Rückspiegel des Steuerrades beigefügt wurde.

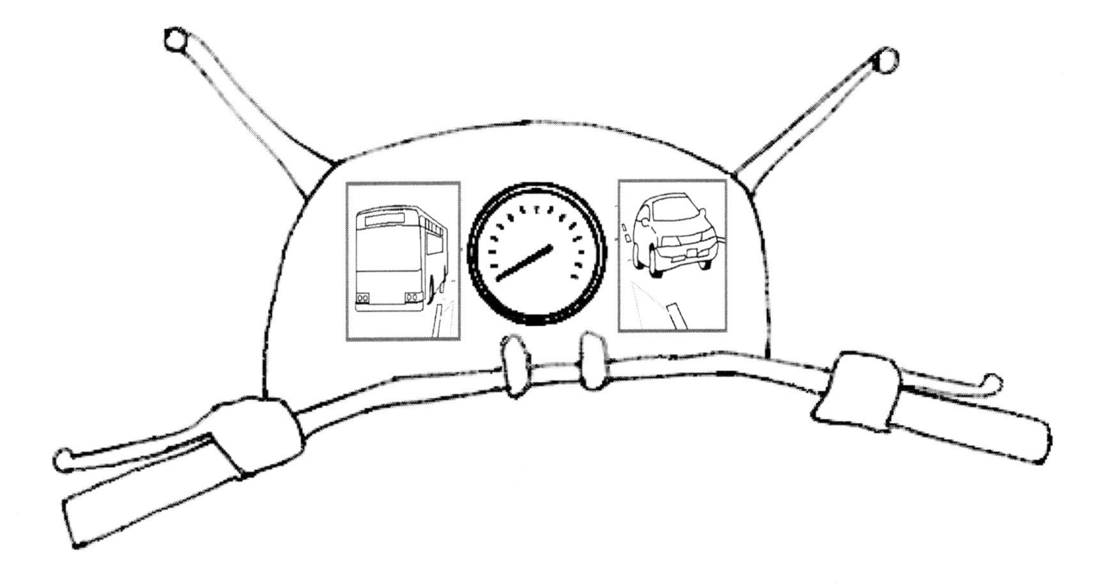

2、

Muster, Übergang, Biegung nach links, Biegung nach rechts, Kreuzung (Illustrationskommentar) der geringen Kollision

(1)

Kontakt des nach links seitigen Musters

Bestätigung der Verhinderung der geringen Kollision durch ein Fahrzeug, das ein Motorrad einholt, das den Kurs ändert

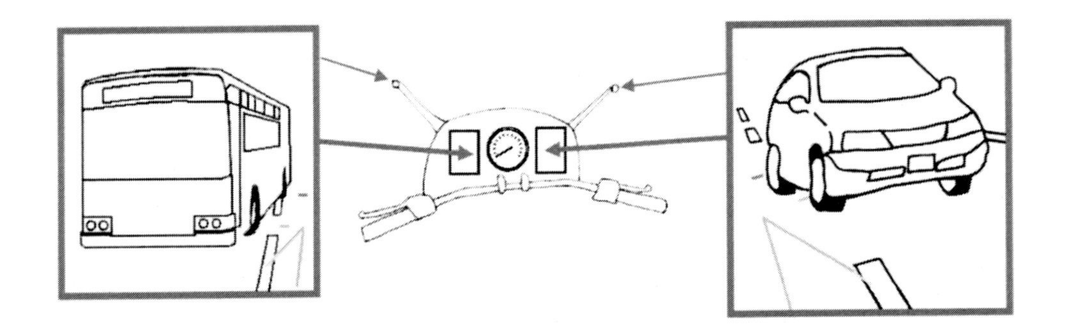

Verpackung wird an beiden Seiten des Tachometers kontrolliert.

Es wird im linken und Recht auf ein Tachometer und ein Tachometer kontrolliert, und es wird hinter dem verlassenen angezeigt.

Es hat zurück im linken und Recht auf das Digitaltachometer kontrolliert, und es wird hinter dem verlassenen angezeigt.

Es hat zurück im linken und direkt in der Meter-Tafel kontrolliert, und es wird hinter dem verlassenen angezeigt.

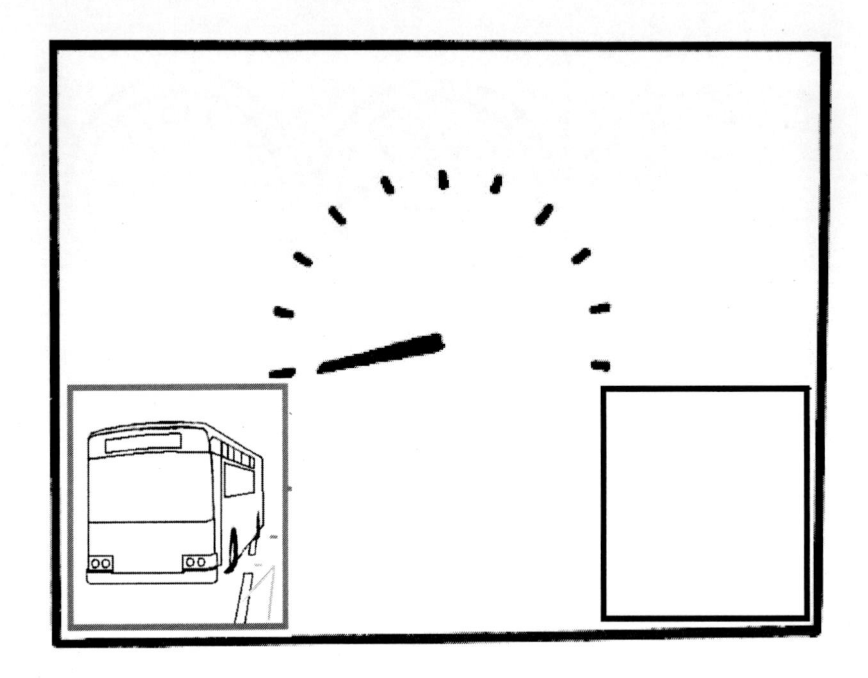

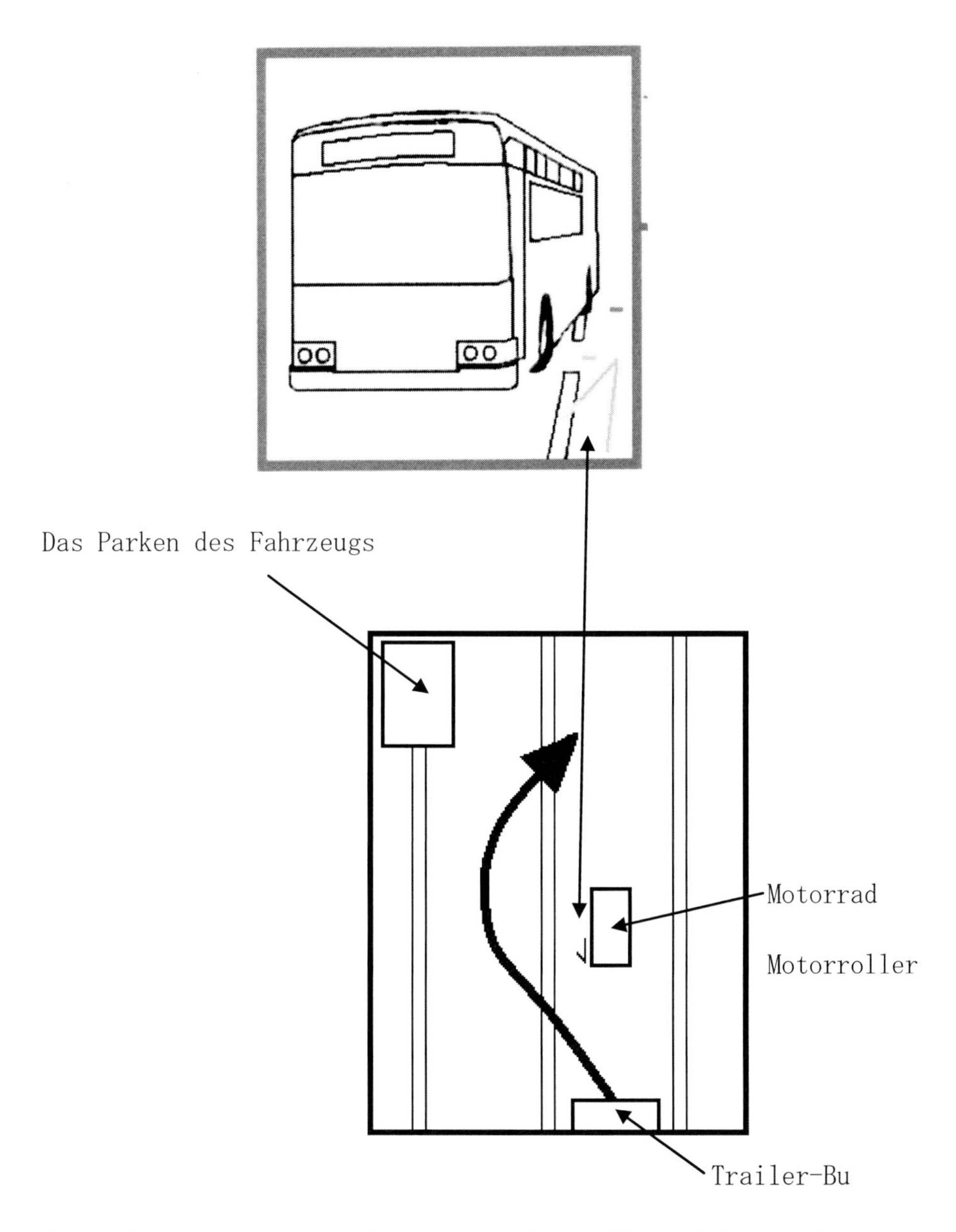

Ich ändere den Kurs zum Recht, ein abgestelltes Fahrzeug in der Vorderseite zu vermeiden, als ein Motortrainer den Kurs vom linken Rücken des zweirädrigen Fahrzeugs geändert hat. Ich sehe Kontakt durch eine Zahl-Linie des Zurückmonitors zur gleichen Zeit voraus.

あとがき

本書のバックモニターの活用範囲は、初版でも解説の如く、著者からの資料は出願当時の平成23年頃からであり、自動車、オートバイ、電動車椅子などの研究著作資料で主に安全性の研究であった。この資料の過程で総合的に著作権企画を行っていたものであり、今後、当初の著作資料に基づいて、追加出版を行う。

また、2005年に創作された統制、ＡＰＩＣによる社会問題、人間並のペット保険登録、ＩＤカード（保険証・診察券）を発行によるシステムの流れの企画書・マニュアルなどの著作権・不正競争防止法による出版を発行予定。

その他、著者は1992年以前から車庫付き中高層建物の研究もなされ、観覧車・エレベータ式などでない昇降方法で、安全・安価にできる構成であり、その構成内容を独創的表現で解説、その解説文献を明細書に引用された場合は、著作権侵害に抵触する可能性がある。これらの技術の解説表現による出版を今後、予定している。

更に医療部門では、アミノ酸などを含んだ癌の治療薬や難病の研究をしており、著者は薬学の専門家でもあることから製薬会社から注目されている特効薬の文献の著作権出版も今後、予定している。

本書のバックモニターに表示される図形ラインは基本パターンであり、パターンには数種類あり、これを文章の著作権で表現すれば「駐車を行う場合、モニターを駐車用の表示に切り替える。Ａパターンの図形ラインは極端に狭い駐車場、Ｂパターンの図形ラインは大型車などである。」これらの構成をイラスト、図形ライン、使用方法を著作権、表現で解説したものであり、更にこの企画の内容は不正競争防止法に鑑みたものである。

オートバイ＆電動車椅子のファイバースコープバックモニター　図形ラインで安全走行

定価（本体 1,500 円＋税）

２０１５年（平成２７年）８月２８日発行

発行所　IDF(INVENTION DEVLOPMENT FEDERATION)
　　　　発明開発連合会®
ﾒｰﾙ 03-3498@idf-0751.com　www.idf-0751.com
電話 03-3498-0751(代)
150-8691 渋谷郵便局私書箱第２５８号
発行人　ましば寿一
著作権企画　IDF 発明開発(連)
Printed in Japan
著者　牧野 真一 ©
（まきのしんいち）

No. 

初版、２０１４年（平成２６年）３月５日発行に記載できなかった原稿の追加発行
